Markus Hadler • Beate Klösch
Stephan Schwarzinger
Markus Schweighart
Rebecca Wardana • David Neil Bird

Surveying Climate-Relevant Behavior

Measurements, Obstacles, and Implications

Markus Hadler
Department of Sociology
University of Graz
Graz, Austria

Beate Klösch
Department of Sociology
University of Graz
Graz, Austria

Stephan Schwarzinger
Department of Sociology
University of Graz
Graz, Austria

Markus Schweighart
Department of Sociology
University of Graz
Graz, Austria

Rebecca Wardana
Department of Sociology
University of Graz
Graz, Austria

David Neil Bird
Joanneum Research
Graz, Austria

ISBN 978-3-030-85795-0   ISBN 978-3-030-85796-7   (eBook)
https://doi.org/10.1007/978-3-030-85796-7

This Palgrave Macmillan imprint is published by the registered company Springer Nature Switzerland AG.
The registered company address is: Gewerbestrasse 11, 6330 Cham, Switzerland

CONTENTS

LIST OF FIGURES

LIST OF TABLES

CHAPTER 1

Introduction

"2019 has been a year of climate disaster," was the headline of a commentary by Geoff Goldrick (2019) in the *Guardian*. It summarizes the many severe events of that year ranging from bush fires in Australia, unprecedented early tropical storms, and heat waves in Europe to record-low sea ice levels in the Arctic. He concludes, "And yet despite all the scientific evidence, all the destruction, all the suffering, all the apocalyptic predictions, and all the strikes and marches, nothing happens. Global CO_2 emissions continue to rise and the world leaders procrastinate."

His dire conclusion notwithstanding, 2019 was also the year that brought a lot of attention to the climate crisis. Fridays for Future protests energized the youth, Greta Thunberg was elected "Person of the Year" by *Time Magazine*, and the media reported frequently about the climate relevance of certain behaviors and the contribution of mobility, consumption, and other factors to greenhouse gas (GHG) emissions. Hopes thus were high that individuals would become more aware of their climate impact and subsequently alter their behavior.

The advent of the coronavirus and the subsequent COVID-19 pandemic brought a sudden end to this attention to the climate. The public's attention switched to the virus and governments reacted to the pandemic

Lead author: Markus Hadler. All results of this book except for the results on lifestyles are based on the project "Measuring CO_2-relevant behaviors in surveys" funded by the Austrian National Bank OeNB (#17892).

1

with various measures, ranging from vague recommendations to strict lockdowns. On the positive side, in terms of climate, the 2020 emissions report (UNEP, 2020) showed that these measures led to a reduction in GHG emissions, especially in the area mobility and transport. However, it is unclear whether these effects will be lasting or if a rebound effect in terms of behaviors will occur. Regardless, the Corona measures made clear that individual behaviors can be altered to an extent that was not anticipated beforehand.

This book focuses on such individual behaviors. Our research addresses the questions of which behaviors are of climate relevance, who is engaging in these behaviors, in which contexts do these behaviors occur, and which individual perceptions and values are related to them. In terms of research methods, we focus on the measurement of climate-relevant behaviors with population surveys. Our goal in this regard is to develop an instrument that allows a valid estimate of an individual's output with as few questions as possible. A concise way of using a questionnaire to estimate an individuals' impact, in turn, provides room for additional questions on attitudes, values, socio-demographics, and so on that are not always grasped in online footprint calculators, smart meters, and similar tools. In this vein, our book also offers a guideline for survey researchers.

We do not stop at measuring climate-relevant behaviors. We also aim to identify factors that shape these behaviors. In doing so, we follow a sociological interpretation and consider behaviors to be shaped by the context as well as by individual characteristics. This part of our endeavor speaks to researchers who are interested in the applicability of different theories and approaches to this specific type of environmental behavior. Furthermore, we also consider the perceived obstacles to acting in a more environmentally friendly way and use this information to highlight possible levers for addressing change.

1.1   Sociology and Climate Research

The relation between the environment and sociology could be described by a Human Exceptionalism Paradigm until the 1970s when Dunlap and Catton proposed the New Environmental Paradigm (Catton Jr & Dunlap, 1978; Harper, 2015). The difference between these two views is that the former considers human development as "exceptional, independent from environmental forces and capable of adapting via cultural change," whereas the New Environmental Paradigm considers humans and the environment to be intertwined. A distinct environmental sociology thus developed only very recently, given the long history of sociology and its roots in the Enlightenment.

The 1970s also marked a change in climate research. Differences and fluctuations in the climate had been discussed for centuries, but only in the 1970s did the scientific opinion start to swing toward agreeing on an increasing temperature trend (see Dunlap & Brulle, 2015). Subsequently, the Intergovernmental Panel on Climate Change (IPCC) and other institutions were funded to monitor the development of the environment and the climate (see Hironaka, 2014). The IPCC itself publishes reports on the latest scientific findings. The discourse was initially dominated by the natural sciences. The inclusion of social sciences started with an increasing consideration of economics and psychology, which resulted in an emphasis on "systems" and a "methodological individualism" approach when considering human actions.

More recently, sociological views have gained importance as well. These views are discussed in detail in the edited volume of Dunlap and Brulle (2015), which summarizes the thoughts and ideas of the American Sociological Association's taskforce on climate change. Sociology, in their view, needs to discuss the social and political roots of climate change, address its possible consequences for societies, and go beyond the apolitical depiction of climate change measures. In this view, GHG emissions are not equal in terms of their consequences. While reductions due to more efficient coal plants appear to be the same on a GHG scoresheet as reductions due to windmills, only the latter have a greater sustainability potential and can initiate societal transformations. Furthermore, sociology should also consider social justice aspects. GHG reductions by limiting the living space of less privileged groups may result in the same reduction of emissions as limiting international flights for individuals who are better off. Yet, the justice aspects would not be the same.

Brulle and Dunlap (2015) also emphasize that sociological approaches go beyond a singular focus on individuals. For example, geographical characteristics; political, historical, and institutional frameworks; and structural contexts must be considered as well. In some cases, these contextual situations can be grasped by studying country specifics such as particular infrastructure, laws and regulations, financial incentives, the price level of various behavioral offerings, the state of the national economy, and cultural and social norms (Kollmuss & Agyeman, 2002; Stern, 2000). International comparative studies are particularly interesting in this regard, because that possible differences can also be attributed to these factors among all the other potentially influencing factors. This can be seen in a study on environmentally friendly private and public behavior (Hadler & Haller, 2011, 2013), which showed that private behaviors are much more common in countries when appropriate means are provided, whereas public behavior is less context dependent.

In sum, individual actions take place within a social framework. Sociological approaches aim to counteract an over-emphasis on the individual and rational actor models by also considering aspects such as the influence of class, status, conspicuous consumption, and also routines and habits (Ehrhardt-Martinez et al., 2015). Furthermore, steering effects can also emanate from a moral or value level, which is often not fully recognized in rational choice models. For example, certain traditional religious beliefs can have positive effects on environmentally relevant behavior—such as the non-consumption of the GHG problematic beef in India or the limited use of technology by Orthodox Jews on the Sabbath. Climate-relevant attitudes, in addition, not only are important for associated environmental behaviors but also influence political opinions. Attitudes toward GHG emissions, for example, turned out to be a relevant issue when it comes to political polarization in the United States (Dunlap et al., 2016).

Finally, considering the scope of recommendations, Brulle and Dunlap (2015) criticize a post-political stance of climate change research. Results and studies take the political and social background of a neo-liberal world for granted, and recommendations for solving the climate crises are only expressed within this framework. In this regard, they point to the report "America's Climate Choices" by the National Research Council which acknowledged that population growth and economic growth are the main sources of the climate problem in the United States. Yet, the report also stated that it will not look into adjusting these two factors, as they are beyond political acceptability. According to Dunlap and Brulle (2015), this thinking is also engrained in the ICPP reports, which do not consider measures such as birth control or curbing economic growth. Taken-for-granted views and aspects, however, were shattered in the final phase of our research, when the COVID-19 crisis led to unprecedented restrictions on social life. We thus will try to include some insights from the current research on this topic in our concluding chapter.

1.2   CLIMATE-RELEVANT BEHAVIOR AND IMPACT

The main title of this book is "Surveying Climate-Relevant Behavior."
However, what is a climate-relevant behavior? Using Stern's (2000) typol-
ogy of environmental behaviors, "relevance" refers to Stern's definition of
impact, which is "the extent to which it changes the availability of materials or energy from the environment or alters the structure and dynamics of
ecosystems or the biosphere" (Stern, 2000, p. 408). Furthermore, we are
focusing on private-sphere behavior and leave out the dimension of public
behavior. Private behaviors have a direct impact on the environment, while
public behaviors such as protesting and other forms of activism have
mostly indirect effects (Stern, 2000, p. 409).

The impact of behaviors, however, depends strongly on the context.
The amount of emissions from using an electric stove, for example,
depends on the production of the electricity. Comparing the United States
and Europe shows that in the United States 63% of the electricity is produced from fossil fuels, 20% from nuclear energy, and 18% from renewable
sources (EIA, 2021). In contrast, the figures for the EU in 2018 are 40%
fossil fuels, 26% nuclear power, and 33% renewable resources (Eurostat,
2021). On average, an individual cooking a meal in the United States has
a larger GHG impact than an individual cooking the same meal in Europe.
These differences become even more amplified when more energy-
intensive behaviors such as the use of electric cars are considered.

The main geographical focus of this book is Austria and Europe. The
national contexts hence are "advanced industrial nations"—countries that
have a large overall per-capita GHG impact and thus are of particular relevance when addressing climate change. According to the UNEP emission
report (2020, p. XXV), the richest 1% of the world population produce
twice as many emissions as the poorest 50% emit in total. Furthermore, as
Rosa et al. (2015) point out, global and local inequalities are heightened
as richer nations outsource the environmental burden to poorer countries.
This type of international outsourcing of emissions is happening by shifting the dirty production to developing countries. To address this shift, we
follow a consumption-based approach when assessing an individual's
GHG impact. All emissions are considered that are emitted in the production and transportation of a good or a service, regardless of the place of
production.

As for the overall environmental impact, it is usually assessed at the
societal level using the IPAT formula (with "I" referring to the impact,

"P" to the population size, "A" to affluence in terms of GDP per capita, and "T" to technology) or its sociological version of Stochastic Impacts by Regression on Population, Affluence and Technology (Rosa et al., 2015). Technology, as Rosa et al. (2015, p. 37) point out, refers to "all other things, such as culture, institutional practices, and political processes" and not just to technology. Research was able to identify complex relationships between the variables, uneven exchanges between countries, treadmills in production, and other factors. Yet, the basic insight is that population, production/consumption, and land use are the most important drivers of the societal impact.

We use two indicators for the climate impact. First, greenhouse gas (GHG) measured in CO_2 equivalents is used in most chapters. This measure combines the emissions from various GHGs based on their global-warming potential (UNEP, 2020). Furthermore, one chapter considers the energy consumption of respondents, which opens another interpretation. Altering the energy production from fossil fuels to renewable resources results in reduced GHG emissions for certain behaviors. Yet, the energy demand remains the same and prevents using energy for other purposes.

Finally, we need to point out that we focus on the individually "caused" parts of the emissions, as discussed in detail in Chap. 3. Considering the main sources of emissions often points to specific industries and economic sectors, whereas the emissions of a single individuum are small in comparison (Perrow & Pulver, 2015). In Austria, to name an example, around 50% of the industrial sector's CO_2 emissions and around 10% of the total CO_2 emissions are produced by a single steel company.[1] Yet, the UNEP emission report (2020, p. 62) estimates that household consumption accounts for around two-thirds of global GHG emissions, as households consume the goods produced by the industry.

---

[1] https://kurier.at/wirtschaft/bei-der-voestalpine-droht-in-oesterreich-kurzar-beit/400735461 Accessed: Apr 7, 2021.

1.3   MEASURING CLIMATE-RELEVANT BEHAVIOR
IN SURVEYS

Our book aims to improve the use of surveys in measuring climate-relevant behaviors. This goal raises the underlying question of which advantages survey data can offer compared to "hard" data sources such as national statistics on GHG emissions or data derived from smart devices.

Data derived from the use of "smart" technologies such as smart meters, mobile-phones, GPS data, and so on capture only the behavior of individuals who use these technologies. In contrast, survey data can be collected in a representative manner for the general population. Furthermore, the consideration of different sectors of social life such as work, leisure, and consumption would require data from numerous "smart" devices. These data would have to be captured and combined at the individual level, which raises data protection concerns and issues with the recruitment of participants, who use "smart" devices in all areas. Finally, "smart" devices cannot capture the underlying intentions, values, and beliefs of their users. They also cannot determine which part of a purchase is intended for which household member or whose behavior is to be assessed in terms of energy use and emissions. Several explanatory variables would therefore still have to be collected in addition to achieve a "holistic picture."

Aggregated data such as national inventories of emissions can be used for an overall per-capita view, but do not allow one to make inferences about the behavior of specific individuals and groups. National averages at a per-capita level, for example, assign each individual an equal portion of the national figure. Differences between groups and specific usage patterns, hence, cannot be grasped. Survey research can be useful in this regard as it allows the consideration of the consumption pattern of different individuals and social groups. This information, in turn, can be used to study the link between social structure, attitudes, and the GHG impact. Yet, survey research has its own caveats such as relying on reported behaviors, sampling biases, and problems in the data collection process. We aim to minimize some of these problems and to develop a brief instrument that captures most of an individual's GHG emissions and leaves room for other survey questions.

Our approach of surveying climate-relevant behavior starts with identifying the GHG emissions-relevant areas of social life and the development of related survey questions. This process also includes a test of the validity of these questions before they are used to estimate the GHG impact of a respondent. Given that the respondents live in different areas and regions, our approach also considers external characteristics. The development of the instrument and the validation was conducted in Austria in 2019. Yet, the final chapter also provides an outlook at the European level.

In sum, the measurement and explanation of environmental behaviors based on surveys will provide insights regarding the presence of certain behaviors and attitudes among different social groups. We can analyze which groups pollute a lot and whether or not there is connection to environmental attitudes. It will also allow us to discover gaps between impact, intention, and attitudes and to make recommendations at this level.

1.4   Research Team and Content of This Book

The results presented in this book are related to the involvement of the authors in various research endeavors. The main underlying project is the study "Measuring the CO_2 Impact Using Survey Research," which was funded by the Austrian National Bank OeNB (#17892). The goal of this project was to develop a survey that is able to capture the GHG emissions of respondents in a reliable and valid manner. The survey data is available at the Austrian Social Science Data Archive (Hadler et al., 2021). Markus Hadler was the principal investigator and all other authors were involved in one form or another. Furthermore, Stephan Schwarzinger and Neil Bird were part of the ECHOES project team, a H2020 funded project (https://echoes-project.eu/) that included a survey of Europeans on climate-relevant behaviors (Reichl et al., 2019). Finally, Markus Hadler and Markus Schweighart also led the development of the 2020 "Environmental Attitudes and Behaviour" questionnaire of the International Social Survey Programme (ISSP; www.issp.org). The attitudinal questions used in the main survey were taken from the ISSP modules. In turn, results from the study presented in this book informed the development of the climate-relevant behavior questions in the 2020 ISSP questionnaire.

Chapter 2 addresses the question of measuring environmental attitudes and behaviors. It points out that the focus in measuring environmental behavior is often on items that are associated with the respondents' intention to do something "good" for the environment. These are often symbolically important behaviors such as turning off lights or recycling. With the growing importance of climate change, behaviors that involve the emission of large amounts of GHG are now being surveyed more frequently. Chapter 2 provides an overview of the scales used so far to measure environmental behavior, in particular emissions-related scales, as well as surveys that included questions on this topic. A look into previous studies indicates that intention-based behavior is explained to a greater extent by attitudes, while impact-based behavior is more closely linked to socio-demographic factors such as income. Considering impact-oriented environmental behavior in an analysis makes it possible to explore the interaction of socio-demographic characteristics, attitudes, behavior, and environmental consequences in different contexts.

Chapter 3 considers different areas of social life, discusses the emissions that are associated with these areas, and shows how the emissions from specific behaviors can be estimated. It starts with a top-down estimate of the consumption-based emissions by life-area. Emissions are organized into segments that may be easily reduced by changing the behavior of an individual and those segments that are fundamental aspects of our society. The latter (e.g., building construction) form a base emission that each member of society inherits that cannot be altered easily by individual behaviors and hence are also not the focal point of this book. Once the emission segments are defined, the remainder of this chapter discusses how to estimate the GHG output and the energy demand of a respondent. There is a trade-off between accuracy and level of detail, and the need to combine bottom-up survey results with the top-down national emissions inventory. In addition, for some segments or items, there may be data limitations (i.e., lack of data). The selection of indicator items and methods to overcome these problems, so that a reasonable accurate estimate of GHG output given survey limitations is achieved.

Subsequently, Chap. 4 deals with the development and validation of our survey. It covers the most important methodological aspects of the underlying study and describes the selection and validation of questions for measuring emission-related behavior. Survey questions are introduced for the main GHG-relevant segments identified in the previous chapter. We also consider alternatives for some questions, compare their reliability and validity, and point to the most suitable versions. To this purpose, the suitability of the questions for measuring impact-relevant behavior is first discussed. Secondly, a comparison is made with available validation criteria that have been collected for this purpose and appropriate questions are selected on the basis of this empirical evidence. Furthermore, this chapter also discusses our methodological approach and the sampling of our main survey in detail.

Chapter 5 presents the results of our main survey. First, a descriptive overview of the emissions of the Austrian population as well as those of our respondents is provided. Through the variety of questions added in the survey, it is possible to get a detailed insight into how Austrians behave in those areas that are actually relevant to emissions. The findings show that the area of mobility—with an emphasis on annual car usage and flights—as well as meat consumption account for around 50% of the annually produced emissions. Chapter 5 continues to identify the most relevant questions for measuring the GHG impact of an individual, that is, to capture the largest possible amount of GHG emissions with the smallest number of items. This attempt results in a selection of five items, which are able to capture more than three quarters of the emissions. The concluding analysis of factors that shape these emissions shows that socio-demographic variables are much more important than attitudes and values.

The previous chapter focused on the total emissions. Chapter 6 considers a multidimensionality of consumption in the form of different energy demands based on the lifestyles of Austrians. The backdrop for this chapter is that existing research demonstrated weak relationships between environment-related attitudes and the overall environmental impact. It proposes that a look at more specific behaviors and their joint occurrence in specific lifestyles is more promising. The analysis identifies five lifestyles based on energy demand in the six areas of social life (housing, mobility, consumption of goods, diet, leisure activities, and information). It concludes that the selection of policy measures must consider the differences between these Energy Lifestyles.

Chapter 7 turns to the subjectively perceived obstacles to lowering one's GHG emissions. The starting point is that many individuals seem to have difficulties adopting environmentally friendly behavior, despite having a strong environmental awareness. This phenomenon is known as the Value-Action Gap. The chapter tests a model that assumes a linear relationship between attitude, intention, and behavior in environmentally related low-cost behaviors—in this case consumption and mobility behavior. Previous studies suggest that discrepancies can occur between personal environmental attitudes, the resulting intention, and the subsequent environmental behavior. In the first part of this chapter, quantitative analyses are used to determine which groups are particularly likely to exhibit a gap between these three factors. Second, these groups are examined in more detail in 15 qualitative interviews regarding the reasons for these discrepancies as well as the desired policy solutions.

The final chapter summarizes the main insights from the previous chapters and tests whether our approach is also applicable at an international level. Our international outlook remains limited to the European context and data from the ECHOES project mentioned above. Initially, we had planned to use data from the International Social Survey Programme, which covers countries across the world. However, its 2020 surveys on environmental attitudes and behaviors were delayed due to the COVID-19 pandemic. Our international outlook, nevertheless, shows that the proposed items in Chap. 5, which capture more than three quarters of the Austrian GHG emissions, also work well in other European countries. Similarly, the lifestyles identified in Chap. 6 are also present in other societies. Correlating the scope of these lifestyles and the explanatory power of our approach with different national characteristics reveals various influences of political institutions, societal affluence, environmental degradation, social demographics, and other national characteristics. The final chapter closes with a look at the impact of the COVID-19 pandemic. As pointed out in the beginning of this introduction, it had a huge impact on GHG emissions (UNEP, 2020). We will present some results based on data collected during the COVID-19 crisis and discuss the influence of this pandemic on environmental attitudes and behaviors. Results suggest that concerns about COVID-19 also affected individual willingness to act for the environment.

REFERENCES

Catton, W. R., Jr., & Dunlap, R. E. (1978). Environmental sociology: A new paradigm. *The American Sociologist*, 41–49.

Dunlap, R. E., & Brulle, R. J. (Eds.). (2015). *Climate change and society: Sociological perspectives.* Oxford University Press.

Dunlap, R. E., McCright, A. M., & Yarosh, J. H. (2016). The political divide on climate change: Partisan polarization widens in the US. *Environment: Science and Policy for Sustainable Development, 58*(5), 4–23.

Ehrhardt-Martinez, K., Rudel, T. K., Norgaard, K. M., & Broadbent, J. (2015). Mitigating climate change. *Climate change and society: Sociological perspectives,* 199–234.

EIA, U.S. Energy Information Administration. (2021). Retrieved April 6, 2021, from https://www.eia.gov/tools/faqs/faq.php?id=427&t=3

Eurostat. (2021). Retrieved April 6, 2021, from https://ec.europa.eu/eurostat/cache/infographs/energy/bloc-3b.html#:~:text=40%20%25%20of%20the%20electricity%20consumed,power%20stations%20burning%20fossil%20fuels&text=Among%20the%20renewable%20energy%20sources,solar%20power%20(4%20%25)

Goldrick, G. (2019). *2019 has been a year of climate disaster. Yet still our leaders procrastinate*. Retrieved April 6, 2021, from https://www.theguardian.com/commentisfree/2019/dec/20/2019-has-been-a-year-of-climate-disaster-yet-still-our-leaders-procrastinate

Hadler, M., & Haller, M. (2011). Global activism and nationally driven recycling: The influence of world society and national contexts on public and private environmental behavior. *International Sociology, 26*(3), 315–345.

Hadler, M., & Haller, M. (2013). A shift from public to private environmental behavior: Findings from Hadler and Haller (2011) revisited and extended. *International Sociology, 28*(4), 484–489.

Hadler, M., Schweighart, M., & Wardana, R. (2021). *OeNB CO2-relevant environmental behavior*. Data will be available for free at the Austrian Social Science Data Archive. www.aussda.at; https://doi.org/10.11587/WQGMKY

Harper, C. (2015). *Environment and society: Human perspectives on environmental issues (2-downloads)*. Routledge.

Hironaka, A. (2014). *Greening the globe*. Cambridge University Press.

Kollmuss, A., & Agyeman, J. (2002). Mind the gap: Why do people act environmentally and what are the barriers to pro-environmental behavior? *Environmental Education Research, 8*(3), 239–260.

Perrow, C., & Pulver, S. (2015). Organizations and markets. *Climate change and society: Sociological perspectives*, 61–92.

Reichl, J., Cohen, J., Kollmann, A., Azarova, V., Klöckner, C., Royrvik, J., Vesely, S., Carrus, G., Panno, A., Tiberio, L., Fritsche, I., Masson, T., Chokrai, P., Lettmayer, G., Schwarzinger, S., & Bird, N. (2019). *International survey of the ECHOES project*. Dataset. Zenodo. https://doi.org/10.5281/zenodo.3524917.

Rosa, E. A., Rudel, T. K., York, R., Jorgenson, A. K., & Dietz, T. (2015). The human (anthropogenic) driving forces of global climate change. *Climate change and society: Sociological perspectives, 2*, 32–60.

Stern, P. C. (2000). Toward a coherent theory of environmentally significant behavior. *Journal of Social Issues, 56*(3), 407–424.

United Nations Environment Programme. (2020). *Emissions Gap Report 2020*. Nairobi. https://www.unep.org/emissions-gap-report-2020

CHAPTER 2

Measuring Environmental Attitudes and Behaviors

This chapter[1] discusses the basis of our approach, situating it in the existing research on environmental attitudes and behaviors and presenting related scales and surveys. First, the concepts of environmental attitudes and behaviors are discussed with a special emphasis on behavior that has an impact on the environment. Subsequently, the key empirical findings on the factors shaping these attitudes and behaviors will be presented. Afterward, we describe the measurement variants of these dimensions. In particular, the focus is on the measurement of environmental attitudes and behaviors in surveys and on the sub-dimension of emissions-related behavior, which has rarely been included in social science surveys. Finally, obstacles to the inclusion of emission-relevant questions in the survey context are identified.

2.1 ENVIRONMENTAL ATTITUDES AND BEHAVIORS

There is not one single definition of environmental attitudes that is generally accepted. Some approaches see environmental attitudes as a unidimensional construct. Gifford and Sussman (2012) for example, define environmental attitudes as a "concern for the environment or caring about environmental issues" (p. 65). Their focus is on the emotional attachment to nature and an associated sense of worry. Other approaches emphasize the multidimensional nature of environmental attitudes and often

[1] Lead author: Markus Schweighart.

M. Hadler et al., *Surveying Climate-Relevant Behavior*,
https://doi.org/10.1007/978-3-030-85796-7_2

15

distinguish between the affective (corresponds to emotional involvement, i.e., concern), the cognitive (environmental knowledge), and the conative (behavioral intention) dimensions (Maloney & Ward, 1973; Diekmann & Preisendörfer, 1998). The concept of environmental attitudes is broader in the latter approaches since the level of knowledge and the level of behavioral intentions are also seen as part of the attitudes.

One of the most influential approaches is the work of William Catton and Riley Dunlap (1978). They consider different environmental attitudes as expressions of one underlying worldview and proposed a concept of an environmental sociology, which focuses on the interaction between the environment and society. They argue that no matter how different sociological theories were, they all shared as a common characteristic an anthropocentric worldview—that is, humans were always emphasized as unique beings standing above nature. Because of the great importance of culture, which can consciously be adapted and changed, the belief in the human ability to solve social, technical and thus ecological problems is deeply rooted in this dominant way of thinking. The authors labeled this perspective the Human Exceptionalism Paradigm (HEP) and believed that it is the reason why sociology struggles to deal with the social implications of ecological problems. The authors thus proposed the New Environmental Paradigm (NEP), which emphasizes the inseparable connection of humans with nature and the limits of the physical and biological world.

Alongside environmental attitudes, social research is interested in measuring environmental behavior. Anja Kollmuss and Julian Agyeman (2002) understand environmental behavior as "behavior that strives to minimize the negative impact of one's own actions on the natural environment" (p. 214). It thus comprises a variety of practices, such as measures to reduce energy consumption, waste avoidance, restrictions on the purchasing consumer goods, and conscious choice of transport but also compensation payments for flights or participation in environmental protest campaigns.

A basic distinction can be made between public- and private-sphere environmental behaviors (Stern, 2000). Behaviors that affect the public-sphere range from environmental activism, such as protest and direct actions, to less intense forms, such as participation in petitions or the vote for a green political party. The goal of this type of behavior is to point out problems and to motivate decision makers (in politics and business) and other citizens to act in an ecologically responsible manner. Private-sphere behavior, in contrast, includes personal environment-related behaviors and refers to individual actions with the goal to reduce one's personal impact on the environment or mitigate the negative environmental consequences of one's own lifestyle.

Furthermore, when measuring energy and resource consumption, a distinction is often made between direct and indirect energy use (Benders et al., 2006). Direct energy use refers to the direct energy consumption that is used for heating the living space, electricity, and operating motorized vehicles. Indirect energy use, alternatively, refers to the energy consumption required for the production, distribution, and waste disposal of consumer goods and services.

Brigitta Gatersleben et al. (2002) emphasize that environmentally friendly behavior is often defined by researchers based on popular notions of environmentally significant behavior (pro environmental behavior) rather than on impact. Recycling, for example, is undoubtedly environment-oriented behavior, but compared to other behaviors, it does have a rather small impact in terms of energy consumption or greenhouse gas (GHG) emissions (Wynes & Nicholas, 2017).[2] Andreas Diekmann and Peter Preisendörfer (1998) note that certain behaviors are surveyed because they are understood as cognitive proxies for environmental behavior in general. Turning off the light when leaving a room is here a prominent example. However, in the case of such typical environmental behavior—such as switching off lights, separating waste, or saving water—the actual effect on the environment in terms of energy consumption or emissions is often minor.

Some of these problems arise from the definition of environmental behavior we encountered above. Focusing on behavior that *strives* to reduce the negative impact on nature means that the will to do something good for the environment is more important than the actual consequences. Hence, Paul Stern (2000) emphasizes the important distinction between environmental intent and environmental impact. This fundamental

---

[2] However, one might argue that this depends on the definition of *impact*. This is true, as it is not too difficult to find contexts, in which a lack of recycling leads to severe pollution and accompanying problems. For example, when waste is deposited in landfills, serious problems often arise, not only in terms of the environment, but also in social and political terms. One only has to think of electric waste dumps in African countries, which poison valuable groundwater and at the same time represent a precarious means of existence for many children and families and look for copper (see, e.g., Perkins et al., 2014). But from the perspective applied in this book, namely the GHG emission perspective on impact, recycling is still not that important.

distinction calls for another important definition—the meaning of *negative impact*. Negative impact on the environment can be the pollution of air and water and shrinking natural habitats but also direct interferences in the ecosystem such as hunting and fishing.

Against the background of the urgent problem of climate change, impact in this work refers to the GHGs emitted that are directly and indirectly associated with behavior and energy consumption. Building on this, the differentiation between *general environmental behavior* (intention oriented) and *emission-relevant environmental behavior* (impact oriented) is crucial. General environmental behavior is based more on the intention to do something good for the environment and usually covers many areas of behavior. Emission-relevant environmental behavior concentrates only on those areas in which the highest levels of GHG emissions occur.

2.2   Factors Influencing Environmental Attitudes and Behaviors

Environmental Attitudes

After presenting the basic concepts of environmental attitudes and behaviors, we now turn to the central explanatory factors. Generally, the influence of socio-demographic determinants on different aspects of environmental attitudes is rather weak—with an explained variance of below 20% (Klineberg et al., 1998). However, studies suggest there are significant effects of some socio-demographic characteristics (Dunlap & Jones, 2002; Zelezny et al., 2000; Grønhøj & Ölander, 2007). The results show that gender, age, and education level significantly influence environmental awareness in such a way that younger and better-educated individuals are characterized by a higher level of environmental awareness. An explanation for the gender effect may be found in social gender role models, in which women tend to be socialized to be empathetic and nurturing, whereas men are socialized to be more competitive and materialistic. Greater empathy and a stronger concern for others also affect how we deal with the environment. Furthermore, it has been shown that a liberal political attitude tends to be related to a higher environmental concern (Marquart-Pyatt, 2008; Van Liere & Dunlap, 1981; Samdahl & Robertson, 1989).

Table 2.1 Scales for measuring emission-related environmental behavior

Dimension	Bodenstein et al., 1997	Gatersleben et al., 2002	Armel et al., 2011	Bohunovsky et al., 2011	Huddart Kennedy et al., 2015	Markle, 2013 (pro environmental behavior)
Housing	m^2 per capita; type of apartment; type of heating	Heating type; room temperature; inventory of various household appliances (kitchen, sanitary, entertainment); showering and bathing behavior	Frequency of use: laundry, showers (+ duration), elevator, air conditioning, TV (h/day), PC; turn off lights; reduce heating	Inventory and use of electrical appliances; electricity consumption; shower and bathing behavior; heating (system, temperature, ventilation); cooking; lighting; apartment (m^2, year of construction, type); building shell (renovation)	m^2; average energy consumption per m^2	Turn off lights/TV; standby; turn back heating/air conditioning; reduce showering time; washing machine only when full

(continued)

Table 2.1 (continued)

Dimension	Bodenstein et al., 1997	Gatersleben et al., 2002	Armel et al., 2011	Bohunovsky et al., 2011	Hudart Kennedy et al., 2015	Markle, 2013 (pro environmental behavior)
Mobility	Km per year for different means of transport; vehicle ownership	Annual mileage by car; number of cars; annual use of public transport; Vacation: Long-haul flights, means of transport	Check tire pressure; speed on freeway, type of car, trips/week with car, carpool, bus, train	Number and type of vehicles in the household; use of vehicles (km per year passenger cars; fuel consumption; fuel); frequency of use of public transport; frequency of air travel	Car km per week; fuel consumption; air traffic: flights and average distance	Frequency last year: carpool/public transport/walking or cycling instead of car use
Diet	5-level scale (organic, little packaging, seasonal, regional)		Quantity consumed per month: milk, cheese, butter, meat (poultry, beef, pork), fish, seafood, various fruits and vegetables; Regional? Organic?			Reduction of meat consumption (beef, pork, poultry) last year

Waste	4-level scale: recycling of different materials		Amount of waste per week (full shopping bags); frequency (in %): reuse (paper, shopping bags, cups); loose fruit; frequency of consumption: plastic bottles, cans, sweets; Buying CO_2 credits		Frequency of waste separation	
Consumerism	Clothing: wearing time; second-hand portion	Estimate of annual energy consumption (gigajoules); focus on existing inventory	10 dimensions	9 dimensions	Partial CO_2 balance (4 dimensions)	42 items, 4 factors
	5 dimensions are standardized, weighted (housing 8, mobility 10, nutrition 8, clothing 3, recycling 3) and summed up					

As for the cross-national comparison of environmental attitudes, the most studied topics are the influences of affluence and of post-materialistic values. Based on Ronald Inglehart's (1997) approach, in which post-material values lead to environmental concern, a debate arose in the scientific field. Facing empirical evidence, Inglehart later included severe environmental degradation as an explanatory factor, which is often found in poorer countries as a source of environmental concern (Inglehart, 1997). The mutual influences of levels of affluence, post-materialism, and objective environmental pollution have not been resolved sufficiently. The studies carried out for this purpose, which came to partly contradictory results, used very different data, models, and measurements (Marquart-Pyatt, 2008). Results based on multilevel analysis show that environmental concern is higher in poorer countries but that within countries, more affluent people are slightly more concerned about the environment (Fairbrother, 2013).

General Environmental Behavior

According to an influential paper by Kollmuss and Agyeman (2002), a general distinction should be made between internal and external factors when explaining environmental behaviors. External, or contextual, factors include infrastructure; the political, legal, and economic situation; and social and cultural factors. As for internal factors, the most influential theoretical approaches come from psychology. Stern (2011) further distinguishes psychological theories in approaches that are based on individualistic motives and those that place more emphasis on social norms. The former are variations of rational choice concepts, such as the Theory of Planned Behavior (Ajzen, 1991), which focuses on behavioral costs and on personal utility. In contrast, the value-belief-norm theory (Stern et al., 1999) and the norm-activation model (Schwartz, 1977) point out that environmental behavior is motivated pro-socially. Bamberg and Möser (2007) show in a meta-analysis that individualistic motives (e.g., attitudes, problem awareness, and perceived behavioral control) as well as social norms are significant independent predictors of environmental behavior. Different studies supplement these psychologically grounded determinants with contextual aspects, such as behavioral abilities (Gatersleben et al., 2002), opportunity structures (Hadler & Haller, 2011), behavioral costs (Diekmann & Preisendörfer, 1998), dwelling characteristics (Perkins et al., 2014), and different settings (at home, at work, etc.; Bratt et al.,

2015). Additionally, some macro-level factors such as wealth, urbanization, and level of post-materialism (Pisano & Lubell, 2017) also turn out to be influential in this regard.

Moreover, some studies highlight the influence of socio-demographic attributes such as age, sex, and income, and of attitudes and values (Diekmann & Jann, 2000; Dunlap et al., 2000; Huddart Kennedy et al., 2015; Gatersleben et al., 2002; Lenzen et al., 2004; Pisano & Lubell, 2017). Results indicate that women show more and stronger environmentally friendly behavior. Findings on age are less consistent as some works have found more environmentally friendly behaviors amongst older people, whilst others found no such effect. A higher income does not affect general environmental behavior (but it comes along with a higher energy demand and GHG emissions, as we will see later). Finally, Bratt et al. (2015) show that some private environmental behaviors (i.e., home-based actions) are strongly correlated with impression management scales. Individuals who are interested in impressing their counterparts show stronger environmentally friendly behaviors.

Emission-Related Environmental Behavior

"Emission relevance" refers to a specific type of environmental behavior. However, studies have found that a single socio-demographic characteristic has the greatest influence here, namely income. People with higher incomes show increased levels of emissions based on their behavior (Huddart Kennedy et al., 2015; Csutora, 2012). The reasons for this are certainly that many individual emission-related behaviors are highly routine actions (diet, commuting), that they are subject to restrictions that limit one's choices (housing), or that they fall into domains with strong emotions involved, such as mobility, traveling, or diet. The latter examples, while also emissions-related aspects of life, are even more so resources or vehicles for constructing self-identity—and they are a form of social communication (Ehrhardt-Martinez et al., 2015).

Stern (2000) also points out that many environmentally relevant behaviors are part of personal routines and/or are subject to severe restrictions such as lack of infrastructure or financial resources. Based on the idea that behavior is influenced by attitudes, on the one hand, and by context, on the other hand, Stern hypothesizes that with the increasing environmental impact of a behavior, the dependence on attitudes decreases. This is similar to the high-cost hypothesis (Diekmann & Preisendörfer, 1998), which postulates that situations involving high behavioral costs (such as choice of transport) must be clearly distinguished from those involving low behavioral costs (e.g., recycling). The influences of attitudes and values can therefore only be effective in the area of low behavioral costs. The influence of social networks (offline and online) on emission-relevant behavior has also been determined, for example, using the example of nutrition (Christakis & Fowler, 2007).

Altogether, contextual factors and basic conditions explain more of the variance in energy-relevant behavior than do individual features (Newton & Meyer, 2012; Tabi, 2013). Within individual determinants, sociodemographics are stronger predictors for actual consumption (Poortinga et al., 2004), while attitudes are more important when it comes to changing behaviors (Abrahamse & Steg, 2009). It is evident that a strictly individualistic approach is insufficient in explaining environmental behavior since structural effects and contextual conditions are important. Geographical conditions; political, historical, and institutional frameworks; and structural contexts have to be considered as well. In some cases, these contextual situations can be realized by studying country specifics. In particular, infrastructure, laws and regulations, financial incentives, the price level of various behavioral offerings, the state of the national economy, and cultural and social norms should be examined, following established suggestions (Kollmuss & Agyeman, 2002; Stern, 2000). International comparative studies are particularly interesting and challenging in this regard as possible differences can also be attributed to these factors along with all the other potentially influencing factors.

This brief summary shows the complexity of emission-related behaviors and general environmental behaviors. Even more, the relation of these two variants of environmental behavior to environmental attitudes also differs. Environmentally friendly attitudes correlate with general environmental behavior, whereas there is no impact on energy demand and emission-based measures (Huddart Kennedy et al., 2015). This moderate correlation notwithstanding, gaps between environmental attitudes and general behaviors—known as the value-action gap (Blake, 1999)—and between environmental behaviors and the actual ecological consequences of actions—known as behavior-impact gap (Csutora, 2012)—occur frequently. We will address these gaps in Chap. 7.

2.3   MEASURING ENVIRONMENTAL ATTITUDES

In empirically oriented sociology, the study of environmental issues dates back to the 1970s. Since then, dealing with environment-related attitudes has become the basis of empirical environmental social science research (Huddart Kennedy et al., 2015), and measuring attitudes, in conjunction with socio-demographic characteristics, accounts for a large proportion of the research conducted to date (Dunlap et al., 2000; Stern, 2000; Knight & Messer, 2012).

A pioneering piece of research was Michael Maloney and Michael Ward's (1973) psychological study, which used 130 questions to investigate verbal willingness to act ecologically, emotional involvement with environmental problems and environmental knowledge. It thus covered all aspects of the tripartite attitude measurement (see Best, 2011). Subsequently, a variant of this measurement tradition gained great importance in the German-speaking part of the world—the scale for measuring general environmental behavior by Andreas Diekmann and Peter Preisendörfer (1998). Recently, criticism of this tripartite division has increased, and reductions or differentiations have been proposed and developed (Schaffrin, 2011). For example, the cognitive dimension is said to differ from the other dimensions and should rather be treated as their precondition (Bord et al., 2000).

Based on the original NEP concept mentioned before, Riley Dunlap together with Kent van Liere (1978) developed an enhanced item set to measure attitude dimensions, which they called the New Environmental Paradigm. The authors claim that the ecological worldview is one-dimensional and measurable with their scale. An overview of the frequent reuse of this scale and its variants can be found in Lucy Hawcroft and Taciano Milfont (2010). The most prominent evolution of the original scale was done by the original authors themselves together with Angela Mertig and Robert Emmet Jones (Dunlap et al., 2000). This new scale adds further facets of an ecological worldview and comprises a set of 15 items representing both pro and counter positions with a strong internal consistency.

These two measurement traditions of environmental attitudes presented above have dominated the research and evoked a number of follow-up studies. Regardless, several other, completely new measuring instruments have also been developed. The development of ever-new scales has been criticized as it has resulted in a multitude of different operationalizations without proper theoretical foundations (Dunlap & Jones, 2002). An overview of such scales including a critical evaluation is provided by André Schaffrin (2011). Our focus here is on behaviors; hence, we will devote more space to this aspect in the following section.

2.4   Measuring Environmental Behavior

The measurement of environmental behavior has been carried out using a variety of scales consisting of different items, which often only refer to specific aspects of environmentally relevant behavior and do not claim to be able to comprehensively measure it (Markle, 2013). In the following, we will first deal with those measurement variants that are aiming at general environmental behavior and then with those that refer to emission-relevant behavior. The latter are often found in more technically oriented studies that investigate specific behaviors in detail, often relying on technical measurement appliances (e.g., metering electricity consumption or using GPS devices to track mobility). At the same time, often the causal factors, such as attitudes, values, or context, are only poorly captured in these studies and thus do not provide any information about factors that can be helpful to foster behavioral change. At the other end of the spectrum, there are social scientific surveys that are more strongly oriented toward values, attitudes, and behavioral intentions, and thus only allow limited conclusions to be drawn about actual behavior let alone quantifiable emission amounts.

Looking at scales for measuring general environmental behavior, one finds a wide range of behavioral domains. Such scales cover different areas and include hypothetical behaviors, such as a willingness to give up driving or pay higher taxes for environmental protection; public behavior, such as participation in political protests and other forms of activism; and private behavior, such as waste separation or energy conservation (Stern et al., 1999; Kaiser et al., 2003; Iwata, 2004; Mobley et al., 2009; Bratt et al., 2015).

In contrast to these general environmental behavior scales, other instruments have been developed that focus on environmental behavior that has a significant impact on various environmental aspects (see Table 2.1). The behavioral domains of these impact scales are more closely related to the main emission-related sectors. For example, all these scales include the domains of housing and mobility. However, other domains are omitted and some behaviors have even here only minor CO_2 relevance. Some scales, for example, ask for precise information on water consumption, a behavior that is particularly relevant in areas with water shortages. However, in the emissions inventory in Austria, water and its disposal are only of limited relevance, accounting for about 1% of all emissions (Bird et al., 2017).

Furthermore, it appears that some items are not suitable for estimating emissions and the frequency of behavior although they aim to measure the environmental impact. Markle's items (2013) on the frequency of switching off lights, reducing heating temperatures, or shortening showering time provide information on how environmentally conscious a person is trying to behave, but do not tell us how much energy is actually consumed as actual consumption is not captured.

When trying to move away from emissions-related behavior to the total emissions generated by an individual's lifestyle across all areas, one must be careful. In general, one's individual emission level is a practically unmeasurable quantity. Even if, as is often the case, technical aids such as smart meters for power consumption or GPS tracking for car use are employed, the actual emissions remain unclear to a certain extent. If you look at the use of a product from a life-cycle perspective, you must also add the emissions released in the past during the extraction of raw materials, production, and transport. A carefully conducted life-cycle assessment (see Chap. 3), however, is able to approximate the output, which in turn can be used for further evaluations and analyses.

2.5   Survey Programs Considering Environmental Attitudes and Behaviors

An increased focus on the environmental issue can be found in international survey research, such as the *International Social Survey Programme* (*ISSP*), the *World Value Survey*, the *European Social Survey* (*ESS*), and other initiatives as well. Most international surveys, such as ISSP 2010 and ESS 2016, have treated behavioral aspects in a similar way. A closer look at these scales shows that behaviors associated with only relatively low CO_2 emissions, such as waste separation, are prominently surveyed. The reason for the inclusion of such behaviors may partly be that they are seen as indicators of a general positive attitude toward the environment, reflecting environmental awareness. There are six items on environmental personal behavior in the ISSP 2010 measured via statements rated on a four-point scale, ranging from "always" to "never." Two of them ask whether someone makes "a special effort" to recycle and buy organic. The other four statements include the phrase "for environmental reasons" when asking about behaviors. On the one hand, these features are likely to assure a higher level of internal consistency; on the other hand, it shifts the focus toward the intention and thus away from the impact perspective.

Environmental attitudes, alternatively, have been surveyed intensively in these international programs. The ISSP first surveyed a pool of questions on this topic in 1993, which was also the central focus in 2000, 2010, and 2020 (Gesis, 2020). The questionnaires cover aspects such as salience of environmental issues, environmental knowledge, attitudes toward environment, science and nature, the willingness to make trade-offs for the environment, environmental efficacy respectively skepticism, the assessment of dangers of environmental problems, and attitudes toward environmental policies. The sub-dimension of attitudes toward the environment, science, and nature includes items related to the environment/science relationship, attitudes toward economic growth and modern lifestyles. The *World Value Survey* fielded questions on environmental awareness in the second, fourth, and fifth waves of the survey (World Value Survey, 2020), dealing with the willingness to make trade-offs in order to avoid pollution and with an assessment of the dangers of environmental problems for the local and, separately, for the global environment.

The ESS treated behavioral aspects in a similar way. It dealt with the topic of climate change and energy use for the first time in 2016 in Round 8 (ESS, 2020). Regarding attitudes, the questionnaire included items on preferred method of energy production, belief in climate change, an assessment of the consequences and causes of climate change, and preferred method of tackling the causes. The *Eurobarometer* surveys have dealt already several times with the attitudes of the European population toward environmental issues. Since 1999, thematic survey waves have been conducted in this area (environment, biodiversity, climate change, and sustainability) regularly (Eurobarometer, 2020). The specific dimensions that have been surveyed cover a wide range of topics, including the classification of environmental hazards and the personal importance of environmental protection and personal responsibility, intentions, environmental influence on the quality of life, environmental knowledge, and the assessment of the role of political actors when it comes to environmental protection.

Given the increasing importance of climate change, research has started to focus more on this topic. As mentioned before, the ESS fielded a module on this topic in 2016, and the ISSP included some questions on this topic in its 2020 module. Alongside these large international groups, several other initiatives were active. To gain insight into the perceptions of climate change and policy measures to address it, Steentjes et al. (2017) conducted a cross-national survey, called "European Perceptions of Climate Change (EPCC)," in Germany, Norway, England, and France on the perception of the climate crisis as well as the acceptance of different policy measures. The H2020 project ECHOES (https://db.echoes-project.eu/echoes/home) conducted an online survey in 31 European countries (Reichl et al., 2019) that included numerous questions on climate-relevant behaviors and some questions on lifestyle indicators and socio-demographics. Two of the authors of the current book, Neil Bird and Stephan Schwarzinger, were members of the ECHOES team and will use its data.

2.6   Conclusions and Outlook

This chapter provided an overview of the theoretical approaches to environmental attitudes and behaviors, as well as of different scales and survey programs with a focus on this topic. In general, capturing individual behavior is often problematic. Research on environmental behavior has repeatedly shown that self-reported behavior can be subject to strong biases because actions are assessed inaccurately or because respondents try to present themselves in a rather favorable light (Bratt et al., 2015). In addition, it is problematic that some scales, although only one person is interviewed, target the household level (Bohunovsky et al., 2011; Gatersleben et al., 2002; Huddart Kennedy et al., 2015) and are prone to considerable deviations (Grønhøj & Ölander, 2007; Seebauer et al., 2017).

The scales presented in this section have rarely been compared with external criteria. In many cases, validity was checked with reference to the correlation of the dimensions among each other (Armel et al., 2011) or to the correlation with attitude and value scales (Huddart Kennedy et al., 2015; Markle, 2013). Positive exceptions are Gerhard Bodenstein et al. (1997) and Brigitta Gatersleben et al. (2002), who compare the calculated energy consumption with official energy numbers or data from representative studies. However, since in both cases only the mean value of their own samples is compared with average values, specific assessments of the validity of different individual scales and items are not possible.

A validation of self-reports by means of external criteria through observations of the behavioral consequences by the interviewers was carried out with a general environmental behavior scale, called the "General Ecological Behavior Scale," by Florian Kaiser et al. (2003), listed in Table 2.1. Kaiser, together with Jacqueline Frick and Susanne Stoll-Kleemann (2001), was able to show that the subjective behavioral data corresponded well with the observations. However, only those items that could be well observed were specifically selected, such as the possession of clothing made of natural fibers, of energy-saving lamps, or of a solar energy system. Therefore, it is difficult to draw conclusions about the validity of less easily observable behaviors with a greater environmental impact.

We will add to this research by validating our impact scales with different external criteria. The results are presented in Chap. 5. Beforehand, the following Chap. 3 will identify the relevant segments of social life that produce GHG and show how the emissions and energy demand of these behaviors can be estimated.

REFERENCES

Abrahamse, W., & Steg, L. (2009). How do socio-demographic and psychological factors relate to households' direct and indirect energy use and savings? *Journal of Economic Psychology, 30*(5), 711–720.

Ajzen, I. (1991). The theory of planned behavior. *Organizational Behavior and Human Decision Processes, 50*(2), 179–211.

Armel, K. C., Yan, K., Todd, A., & Robinson, T. N. (2011). The Stanford Climate Change Behavior Survey (SCCBS). Assessing greenhouse gas emissions-related behaviors in individuals and populations. *Climatic Change, 109*(3–4), 671–694.

Bamberg, S., & Möser, G. (2007). Twenty years after Hines, Hungerford, and Tomera: A new meta-analysis of psycho-social determinants of pro-environmental behaviour. *Journal of Environmental Psychology, 27*(1), 14–25.

Benders, R. M. J., Kok, R., Moll, H. C., Wiersma, G., & Noorman, K. J. (2006). New approaches for household energy conservation: In search for personal household energy budgets and energy education options. *Energy Policy, 34*, 3612–3622.

Best, H. (2011). Methodische Herausforderungen: Umweltbewusstsein, Feldexperimente und die Analyse umweltbezogener Entscheidungen. In M. Groß (Ed.), *Handbuch Umweltsoziologie*. VS Verlag für Sozialwissenschaften.

Bird, N., Jungmeier, G., Canella, L., Windsperger, B., & Windsperger A. (2017). *Consumption based accounting using LCA.* Stakeholder dialog meeting.

Blake, J. (1999). Overcoming the 'value-action gap' in environmental policy: Tensions between national policy and local experience. *Local Environment,* 4(3), 257–278.

Bodenstein, G., Spiller, A., & Elbers, H. (1997). Strategische Konsumentscheidungen: Langfristige Weichenstellungen für das Umwelthandeln—Ergebnisse einer empirischen Studie. Diskussionsbeiträge des Fachbereichs Wirtschaftswissenschaft der Gerhard-Mercator-Universität—Gesamthochschule—Duisburg.

Bohunovsky, L., Grünberger, S., Frühmann, J., & Hinterberger, F. (2011). *Energieverbrauchsstile. Datenbank zum Energieverbrauch österreichischer Haushalte: Erstellung und empirische Überprüfung.* Sustainable Europe Research Institute.

Bord, R. J., O'connor, R. E., & Fisher, A. (2000). In what sense does the public need to understand global climate change? *Public Understanding of Science,* 9, 205–218.

Bratt, C., Stern, P. C., Matthies, E., & Nenseth, V. (2015). Home, Car Use, and Vacation. *Environment and Behavior,* 47(4), 436–473.

Catton, W. R., Jr., & Dunlap, R. E. (1978). Environmental sociology: A new paradigm. *The American Sociologist,* 13(1), 41–49.

Christakis, N. A., & Fowler, J. H. (2007). The spread of obesity in a large social network over 32 years. *New England Journal of Medicine,* 357(4), 370–379.

Csutora, M. (2012). One more awareness gap? The behaviour–impact gap problem. *Journal of Consumer Policy,* 35(1), 145–163.

Diekmann, A., & Jann, B. (2000). Sind die empirischen Ergebnisse zum Umweltverhalten Artefakte? Ein Beitrag zum Problem der Messung von Umweltverhalten. *Umweltpsychologie,* 4(1), 64–75.

Diekmann, A., & Preisendörfer, P. (1998). Umweltbewußtsein und Umweltverhalten in Low-und High-Cost-Situationen: Eine empirische Überprüfung der Low-Cost-Hypothese. *Zeitschrift für Soziologie,* 27(6), 438–453.

Dunlap, R. E., & Jones, R. E. (2002). Environmental concern: Conceptual and measurement issues. *Handbook of Environmental Sociology,* 3(6), 482–524.

Dunlap, R. E., & Van Liere, K. D. (1978). The 'new environmental paradigm'. *The Journal of Environmental Education,* 9(4), 10–19.

Dunlap, R. E., Van Liere, K. D., Mertig, A. G., & Jones, R. E. (2000). New trends in measuring environmental attitudes: Measuring endorsement of the new ecological paradigm: A revised NEP scale. *Journal of Social Issues,* 56(3), 425–442.

Ehrhardt-Martinez, K., Schor, J. B., Abrahamse, W., Alkon, A., Axsen, J., Brown, K., Shwom, R., Southerton, D., & Wilhite, H. (2015). Consumption and climate change. In R. E. Dunlap & R. J. Brulle (Eds.), *Climate change and society: Sociological perspectives* (pp. 93–126). Oxford University Press.

Eurobarometer. (2020). Eurobarometer surveys on public attitudes to the environment. Retrieved October 8, 2020, from https://ec.europa.eu/environment/eurobarometers_en.htm

European Social Survey. (2020). Data and documentation by round/year. Retrieved October 2, 2020, from https://www.europeansocialsurvey.org/data/round-index.html

Fairbrother, M. (2013). Rich people, poor people, and environmental concern: Evidence across nations and time. *European Sociological Review, 29*(5), 910–922.

Gatersleben, B., Steg, L., & Vlek, C. (2002). Measurement and determinants of environmentally significant consumer behavior. *Environment and Behavior, 34*(3), 335–362.

Gesis. (2020). Environment. Retrieved October 2, 2020, from https://www.gesis.org/en/issp/modules/issp-modules-by-topic/environment

Gifford, R., & Sussman, R. (2012). Environmental attitudes. In S. D. Clayton (Ed.), *The Oxford handbook of environmental and conservation psychology*. Oxford University Press.

Grønhøj, A., & Ölander, F. (2007). A gender perspective on environmentally related family consumption. *Journal of Consumer Behaviour: An International Research Review, 6*(4), 218–235.

Hadler, M., & Haller, M. (2011). Global activism and nationally driven recycling: The influence of world society and national contexts on public and private environmental behavior. *International Sociology, 26*(3), 315–345.

Hawcroft, L. J., & Milfont, T. L. (2010). The use (and abuse) of the new environmental paradigm scale over the last 30 years: A meta-analysis. *Journal of Environmental Psychology, 30*(2), 143–158.

Huddart Kennedy, E., Krahn, H., & Krogman, N. T. (2015). Are we counting what counts? A closer look at environmental concern, pro-environmental behaviour, and carbon footprint. *Local Environment, 20*(2), 220–236.

Inglehart, R. (1997). *Modernization and postmodernization: Cultural, economic, and political change in 43 societies*. Princeton University Press.

Iwata, O. (2004). Some psychological correlates of environmentally responsible behavior. *Social Behavior and Personality, 32*(8), 703–714.

Kaiser, F. G., Frick, J., & Stoll-Kleemann, S. (2001). On the adequacy of self reported behavior. A study of the validity of the general ecological behavior scale. *Diagnostica, 47*, 88–95.

Kaiser, F. G., Doka, G., Hofstetter, P., & Ranney, M. A. (2003). Ecological behavior and its environmental consequences. A life cycle assessment of a self-report measure. *Journal of Environmental Psychology, 23*(1), 11–20.

Klineberg, S. L., McKeever, M., & Rothenbach, B. (1998). Demographic predictors of environmental concern: It does make a difference how it's measured. *Social Science Quarterly, 79*(4), 734–753.

Knight, K. W., & Messer, B. L. (2012). Environmental concern in cross-national perspective: The effects of affluence, environmental degradation, and world society. *Social Science Quarterly, 93*(2), 521–537.

Kollmuss, A., & Agyeman, J. (2002). Mind the gap: Why do people act environmentally and what are the barriers to pro-environmental behavior? *Environmental Education Research, 8*(3), 239–260.

Lenzen, M., Dey, C., & Foran, B. (2004). Energy requirements of Sydney households. *Ecological Economics, 49*(3), 375–399.

Maloney, M. P., & Ward, M. P. (1973). Ecology: Let's hear from the people: An objective scale for the measurement of ecological attitudes and knowledge. *American Psychologist, 28*(7), 583–586.

Markle, G. L. (2013). Pro-environmental behavior: Does it matter how it's measured? Development and validation of the pro-environmental behavior scale (PEBS). *Human Ecology, 41*(6), 905–914.

Marquart-Pyatt, S. T. (2008). Are there similar sources of environmental concern? Comparing industrialized countries. *Social Science Quarterly, 89*(5), 1312–1335.

Mobley, C., Vagias, W. M., & DeWard, S. L. (2009). Exploring additional determinants of environmentally responsible behavior. The influence of environmental literature and environmental attitudes. *Environment and Behavior, 42*(4), 420–447.

Newton, P., & Meyer, D. (2012). The determinants of urban resource consumption. *Environment and Behavior, 44*(1), 107–135.

Perkins, D. N., Drisse, M. N. B., Nxele, T., & Sly, P. D. (2014). E-waste: A global hazard. *Annals of Global Health, 80*(4), 286–295.

Pisano, I., & Lubell, M. (2017). Environmental behavior in cross-national perspective: A multilevel analysis of 30 countries. *Environment and Behavior, 49*(1), 31–58.

Poortinga, W., Steg, L., & Vlek, C. (2004). Values, environmental concern, and environmental behavior: A study into household energy use. *Environment and Behavior, 36*(1), 70–93.

Reichl, J., Cohen, J., Kollmann, A., Azarova, V., Klöckner, C., Royrvik, J., Vesely, S., Carrus, G., Panno, A., Tiberio, L., Fritsche, I., Masson, T., Chokrai, P., Lettmayer, G., Schwarzinger, S., & Bird, N. (2019). *International survey of the ECHOES project.* Dataset. Zenodo. https://doi.org/10.5281/zenodo.3524917

Samdahl, D. M., & Robertson, R. (1989). Social determinants of environmental concern: Specification and test of the model. *Environment and Behavior, 21*(1), 57–81.

Schaffrin, A. (2011). No measure without concept: A critical review on the conceptualization and measurement of environmental concern. *International Review of Social Research, 3*, 11–31.

Schwartz, S. H. (1977). Normative influences on altruism. *Advances in Experimental Social Psychology, 10*, 221–279.

Seebauer, S., Fleiß, J., & Schweighart, M. (2017). A household is not a person: Consistency of pro-environmental behavior in adult couples and the accuracy of proxy-reports. *Environment and Behavior, 49*(6), 603–637.

Steentjes, K., Pidgeon, N., Poortinga, W., Corner, A., Arnold, A., Boehm, G., Mays, C., Poumadère, M., Ruddat, M., Scheer, D., Sonnberger, M., & Tvinnereim, E. (2017). European Perceptions of Climate Change (EPCC): Topline findings of a survey conducted in four European countries in 2016. Cardiff: Cardiff University. https://orca.cardiff.ac.uk/98660/7/EPCC.pdf

Stern, P. C. (2000). New environmental theories: Toward a coherent theory of environmentally significant behavior. *Journal of Social Issues, 56*(3), 407–424.

Stern, P. C. (2011). Contributions of psychology to limiting climate change. *American Psychologist, 66*(4), 303–314.

Stern, P. C., Dietz, T., Abel, T., Guagnano, G. A., & Kalof, L. (1999). A value-belief-norm theory of support for social movements: The case of environmentalism. *Human Ecology Review, 6*(2), 81–97.

Tabi, A. (2013). Does pro-environmental behaviour affect carbon emissions? *Energy Policy, 63*, 972–981.

Van Liere, K. D., & Dunlap, R. E. (1981). Environmental concern: Does it make a difference how it's measured? *Environment and Behavior, 13*(6), 651–676.

World Value Survey. (2020). Retrieved October 2, 2020, from https://www.worldvaluessurvey.org/WVSContents.jsp

Wynes, S., & Nicholas, K. A. (2017). The climate mitigation gap: Education and government recommendations miss the most effective individual actions. *Environmental Research Letters, 12*(7), 074024.

Zelezny, L. C., Chua, P. P., & Aldrich, C. (2000). New ways of thinking about environmentalism: Elaborating on gender differences in environmentalism. *Journal of Social Issues, 56*(3), 443–457.

CHAPTER 3

Life-Areas and How to Estimate Greenhouse Gas Emission Footprints

This chapter[1] now turns to the problem of identifying relevant areas of social life that are climate-relevant and how to estimate the greenhouse gas (GHG) emissions and energy demands of specific behaviors. There are a growing number of footprinting tools, most of which can be found on the internet, which claim to provide a sufficiently good approximation of the CO_2 budget, the ecological footprint or one's own carbon footprint with only a few pieces of information. Paul Padgett and colleagues (2008) were able to show in a comparison of such calculators, most of which are designed for private use, that the results sometimes differ greatly depending on which calculation formulas are used with which behavioral questions and with which conversion factors. For a more recent compilation and rating of 31 footprint calculators, I recommend reading Mulrow et al. (2019).

From a scientific point of view, such tools are sometimes met with criticism because the calculation methods used are not sufficiently transparent (Čuček et al., 2012). However, the general purpose of such calculators is to raise awareness of the emissions associated with one's own behavior. Popular calculators in the Austrian context—and here, of course, the national background is important—are those of the Global Footprint Network (https://www.footprintcalculator.org) and the calculators of the Austrian Federal Ministry for Climate Protection, Environment, Energy, Mobility, Innovation and Technology (https://www.mein-fussabdruck.

---

[1] Lead author: David Neil Bird; "I" throughout this chapter refers to the lead author.

© The Author(s) 2022
M. Hadler et al., *Surveying Climate-Relevant Behavior*,
https://doi.org/10.1007/978-3-030-85796-7_3

37

at/), and JOANNEUM RESEARCH (https://www.lifestylecheck.at/). As one might expect, they do not all give the same estimate. This is because the footprint analyst and website designer must make decisions about what type of emissions to estimate and on which sectors of the society and economy to focus. Due to these decisions, trade-offs are made between accuracy and the level of detail of the emissions inventoried. In addition, there is often a problem that bottom-up emission estimates for individuals may not match average per person emissions in the national inventory.

In this chapter, we will investigate solutions to these questions and present a method for reconciling these problems. We will start by coming from the top-down, discussing what type of emissions should be used for footprinting, and present which sectors of society and the economy should be the focus of the footprint. This will be followed by the formulation of a generalized bottom-up method for estimating and categorizing emissions from everyday life, and the chapter will end by presenting an actual calculation and the tie between the top-down and bottom-up estimates.

3.1   What Type of Emissions to Use, Which Sectors Should Be the Focus

The GHG emission estimates, prepared yearly, in a country's national GHG inventory follow the Intergovernmental Panel on Climate Change (IPCC) guidelines (2006). The inventories are prepared by collating information on activities, such as the combustion of liquid fuels, that occur within the country's borders and multiplying these values by country-specific emission factors. Although the national inventory should capture all GHG emissions that occur within the country, there is no linking of activities. For example, a consumer's decision to eat a piece of cheese includes a chain of linked activities, that is, the cow must be fed, sheltered, and milked; the milk must be processed to cheese, cooled, and packaged; the cheese must be transported to and stored in the supermarket; the consumer needs to transport the purchased cheese to her/his home, where it is cooled; the plate must be washed; and the consumer's waste (packaging and defecation) must be disposed of. All these steps are linked to the consumer's decision, but they appear as separate items in the national inventory. The recognized method for estimating the emissions from the piece of cheese is the life-cycle assessment (LCA). The European Environment Agency defines LCA as *a process of evaluating the effects that a product has on the environment over the entire period of its life thereby increasing resource-use efficiency and decreasing liabilities. It can be used to study the environmental impact of either a product or the function the product is designed to perform. LCA is commonly referred to as a "cradle-to-grave" analysis.*[2]

In addition, some of the emissions may occur outside the nation's boundary and will not appear in the national inventory. The national inventories are production-based, whereas what is need for a proper estimate of a consumer's footprint is a consumption-based inventory (Peters, 2008; Davis & Caldeira, 2010). The EU's Green Deal is facing criticism because it considers production-based emissions only (Fuchs et al., 2020). In the EU-28, the consumption-based inventory for 2016 is estimated as 5.6 Gt CO_2-eq., 27% higher than the production-based inventory (4.4 Gt CO_2-eq.) (Wood et al., 2019). The consumption-based inventory is estimated top-down using multi-regional input-output (MRIO) economic models. The difference between the production-based and consumption-based inventories will vary by country depending on its size and the relative proportion of imports and exports in its economy. For Austria, in 2017, using a bottom-up methodology (Windsperger et al., 2017; Jungmeier et al., 2020), I estimate that the consumption-based emissions (114.7 Mt. CO_2-eq. or 13.1 t CO_2-eq/person) are nearly 40% greater than the national inventory (82.3 Mt. CO_2-eq. or 9.4 t CO_2-eq./person). The consumption-based emissions are given by

$$Consumption_y = Production_y + Imports_y - Exports_y.$$

The consumption-based inventory may be divided into six categories, as shown in Fig. 3.1 (left). Most emissions are caused by the consumption of goods (39%), followed by mobility (33%), housing (12%), food (10%), infrastructure (5%), and waste management (1%). The consumption category can be further subdivided into key consumption items (Fig. 3.1, right). In the diagram below, one can see that steel, food, and textiles all cause more than 1 t CO_2-eq emissions per person. Please note that in Fig. 3.1 (left), food appears as a separate item. A general problem with bottom-up methods is that one cannot calculate the emissions for all items consumed due to limited data and the diminishing contributions of some

---

[2] https://www.eea.europa.eu/help/glossary/eea-glossary/life-cycle-assessment

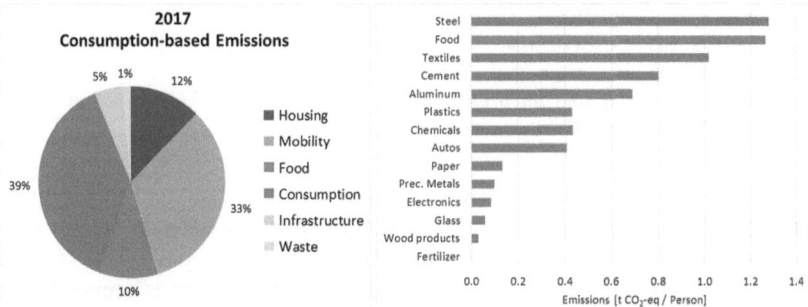

Fig. 3.1 Components of the consumption-based inventory (left) and emissions per person for various goods (right). (Source: Lead author's unpublished self-calculation)

items. There is a trade-off between level of detail and effort to collect the necessary data. This selection of items causes a truncation error, and one should use an IPCC key-category type analysis to identify products that should be included and keep the truncation error to an acceptable minimum (IPCC, 2006). We will visit this problem again when we discuss consumer footprinting.

So now, how does one convert a national consumption-based inventory to a meaningful indicator for personal footprinting? One method is to redo the categorization presented above into the six requirements of everyday life ("life-areas"). We suggest that an individual needs the following:

1. A place to live (housing);
2. To be able to get to work, the grocery store, and so on (mobility);
3. Nutrition (food and part of waste);
4. Clothing, books, furniture, and so on (parts of consumption and waste);
5. Information (parts of housing and infrastructure); and
6. Leisure (part of infrastructure).

Of course, there are gray areas in the categorization. For example, does one classify travel for a vacation as leisure or as mobility? The exact classification is not important as long as it is stated clearly and consistently applied.

The personal footprint suffers from the same problem as the bottom-up consumption-based inventory. The analyst must select which items and services to include in the footprint. Hence, there is this trade-off between level of detail and effort to collect the necessary data. It is relatively easy to collect specific fairly detailed information on mobility and housing. For example, most people know how far they drive in a year, how much time they spend commuting per day, or how much area their dwelling has. It requires more effort by both the analyst and the individual to collect detailed information on nutrition and the consumption of goods and information. For nutrition, the analyst could require the individual to provide information in grams per week on the consumption of types of meats, dairy products, vegetables, grains, and fruits because that is how the analyst needs the information. However, it is a tedious task for the consumer to provide these data via food diaries, so "short-cuts" are probably made. For example, one could define standard eating styles based on the frequency of meat consumption in meals per week. The individual would need to choose which eating style best describes her/his lifestyle.

An important distinction between footprinting and consumption-based inventories is how the information may be used. As discussed in other chapters of this book, the footprint could be used to help an individual change her/his lifestyle or make daily consumer decisions. Hence, another consideration in choosing which items and services form the detailed part of the footprint is which life-areas have influence and which life-areas the individual can change.

For this reason, I suggest a large portion of the individual's footprint is a constant that reflects societal demand and not the individual's demand for certain goods and services. The consumption of items such as steel, cement, aluminum, some plastics, and chemicals fall into this group. These items are used in the construction of buildings and infrastructure in society. The analyst cannot ask the individual: "how much cement did you consume last year?" The individual will answer zero, but it is the largest contributor to consumption in the consumption-based inventory. Not only that, but the individual has a very limited ability to change the amount society consumes.

In this section, I have discussed what type of emissions should be used for footprinting and presented which sectors of society and the economy I believe should be the focus of the footprint. While it is relatively clear what type of emissions should be used (i.e., consumption-based using LCA), there really is no "right" or "wrong" selection. I made my selection of sectors and consumption items from a top-down analysis. The footprint analyst may have other reasons for choosing which sectors and consumption items appear in her/his analysis. However, the reasons should be clearly documented, and the analyst must always keep in mind that there are trade-offs between accuracy and expediency made with these decisions. In the next section, I will look at a general method for estimating emissions from the bottom up.

3.2   A GENERALIZED METHOD FOR ESTIMATING EMISSIONS FROM A SERVICE

The generalized approach for estimating GHG emissions from an activity or service is shown in Eq. 3.1:

$$g_i = s_i \, x\eta_i \, xEF_i \tag{3.1}$$

For item i, g_i are the emissions from an amount of service s_i, which has an efficiency-related indicator η_i, and an emission factor EF_i.

The footprint of a consumer is, then, the sum of the emissions from the individual services.

$$G = \sum_i g_i \tag{3.2}$$

One's choices, as a consumer, affect all three components of Eq. 3.1. One can consume less of the service, use a service from an efficiency system, and/or use a low emission fuel.

Using the generalized equation is the goal of footprint analysis. However, in practice, this equation is simplified as needed to fit the available information and simplify the data collection needs. This is better illustrated with specific examples from each of the life-areas.

Mobility

As a consumer, one requires a specific amount of transportation service (s_i in km). This service is provided by, for example, a car, with an efficiency-related indicator (η_i in MJ per km), and the energy is provided by a fuel with an emission factor (EF_i in g CO_2-eq per MJ). For the individual's daily commute with multiple transportation modes, the generalized equation may need to be modified. The individual may know how much time he or she spends on a bus, in a train or subway, using a bicycle or walking. Each mode has its unique efficiency-related indicator and emission factor. In addition, LCA transportation studies often report CO_2-eq emissions per passenger-km. Hence, the $\eta_i \times EF_i$ term in Eq. 3.1 is replaced by $v_i \times EF_i$ where v_i is the average velocity of the mode of transport.

Housing

The consumer requires a specific amount of floor space to live as a service (s_i in m^2). The energy demand for heat per unit floor space is the efficiency-related indicator (η_i in MJ per m^2) that depends on the building envelope and heating system, and the emission factor (EF_i in g CO_2-eq per MJ) is determined from the type of fuel that is used to provide the energy.

Nutrition

Every person needs a daily amount of food energy. This is supplied by a gamut of food types. Each food type, for example, legumes, provides a certain amount of food energy (s_i in kcal). The food has an efficiency-related indicator (η_i in kg per kcal), and there is an emission factor to produce the required number of legumes (EF_i in g CO_2-eq per kg).

Of course, to build an individual's footprint, it would be nice to know the amounts consumed of many food stuffs, but, without the individual filling out a detailed food consumption log, this is beyond the knowledge of the individual. In addition, the emission factors for different foods within a food category are often similar (e.g., carrots and parsnips). Hence, the data collection may be simplified by assuming standardized dietary categories. For example, an individual may follow a vegan, vegetarian, or pescatarian lifestyle. If the person eats meat, then he/she may be asked their frequency of meat consumption by type from once a week to daily. In addition, emission factors differ for food stuffs depending on whether they come from organic or non-organic production. Hence, for nutrition, it is simpler to apply Eq. 3.1 to patterns of services that specify amounts of food stuffs.

Consumption

Emissions from the consumption of goods come from many sources, and the individual may have limited knowledge about his/her consumption. For example, can you say how many tons of cement or steel you consumed last year? If you do know, could you change your lifestyle to reduce the emissions? The consumption of cement and steel (other than in vehicles) occurs primarily in the construction of buildings and infrastructure. In my experience assessing footprints, I have chosen not to focus on these items but rather allocate them to the individual as a societal "overhead." Instead, I have focused on the consumption of items that defines a person's lifestyle, including clothing, electronics, sporting goods, and vehicles. For example, are you a clotheshorse? An electrophile? A sport's buff? A petrohead? Once aware of the impact of her/his lifestyle, an individual can actively reduce his/her footprint.

Following Eq. 3.1, for these items, the service is the number of items (or fraction thereof) purchased per year. For clothing, this may be the number of shirts, jeans, underwear, shoes, sweaters, and outer garments purchased per year. A less detailed inventory may be quantified by asking the individual if they purchase more or less (as a range) clothing than their colleagues. Nevertheless, the answer must be converted to an amount of clothing (kg) of various types, which the individual purchases based on some assumptions.

For electronic goods, a similar approach can be used. The individual can be asked how frequently they purchase a cell phone or flat screen monitor and whether they purchase new or used products.

This leads to the temporal aspects of the footprint. In LCA, the environmental impacts of a good or service are usually amortized over the lifetime of the good or service, where lifetime may be in terms of time (e.g., years) or service (automobile lifetime in kilometers). For footprints, we are interested in time-average emissions even though there may be a large temporal variation in emissions. For example, if you moved into a newly constructed apartment this year, your instantaneous footprint is dominated by this purchase, and as discussed above, its construction and the emissions thereof form a significant component of one's footprint. However, as I have chosen to assign these as a societal burden, and applying ergodic theory, spatial and time averages are equivalent. In general, I propose that there are the following three types of consumer goods:

1. Items of which an individual has one. These are discarded and replaced after a useful lifetime. They are rarely traded for reuse by another consumer. An example of a type 1 item is one's cell phone.
2. Items of which the individual has many, but only one may be used at one time. New units are purchased regularly but rarely when the old is discarded for reasons other than end-of-life. Hence, the true lifetime of the unit is longer than the useful lifetime, and units are stockpiled. An example of a type 2 item is a pair of jeans.
3. Items of which the individual has one, and it is replaced for reasons other than end-of life, but the item is not discarded. It is traded to a second consumer. Probably the most important type 3 item is an automobile.

For each of these types of consumer goods, I suggest that their footprints are estimated using different temporal methods. This is partly due to their nature but also due to the individual's knowledge of her/his consumption.

For type 1 consumer goods, the annual emissions are calculated using simple amortization. The individual knows fairly well the typical lifetime of the item. Simple amortization is used in LCA. However, simple amortization has its drawbacks (see, e.g., the discussion about emissions from bio-based products in Liptow et al., 2018).

I prefer, and recommend, using the instantaneous annual emissions for type 2 items. The annual emissions are calculated from the annual consumption of the item. For example, the individual is asked how many pairs of jeans he/she bought last year. My recommendation is based on two points. Firstly, the individual has this knowledge, while they may not be able to answer how long a pair of their jeans lasts. Secondly, the purchase of new items may extend the lifetime of the existing items. For example, the new pair of jeans is worn instead of the older pair of jeans, but the older pair is not discarded. Its lifetime is extended even though that may mean that the pair of jeans collects dust in the consumer's closet.

Consumer goods of type 3, in particular automobiles, are a bit of a problem. The conventional simple amortization means that the individual is allocated a fraction of the emissions for the vehicle dependent on the lifetime of the vehicle independent of whether he/she drives a new or used car. This means that there is no footprint surcharge for frequently purchasing a new vehicle, which is not a lifestyle that we wish to promote, from my perspective. If fact, I argue that the purchase of a new car may temporarily create an oversupply of used cars, lowering their price. This may ripple through the used car market until an individual with a vehicle near end-of-life decides that it is now cheaper to purchase a new (to them) used car instead of repairing her/his existing one. As a result, I argue that the purchase of a new car shortens the lifetime of a used car and increases the individual's footprint. Figure 3.2 is the author's modeled impact of frequency of new car purchase. I assume that the individual receives credit for her/his old car when she/he purchases a new one and that vehicles are subject to exponential decay (as a constant percentage of vehicles is lost each year due to accidents, etc.).

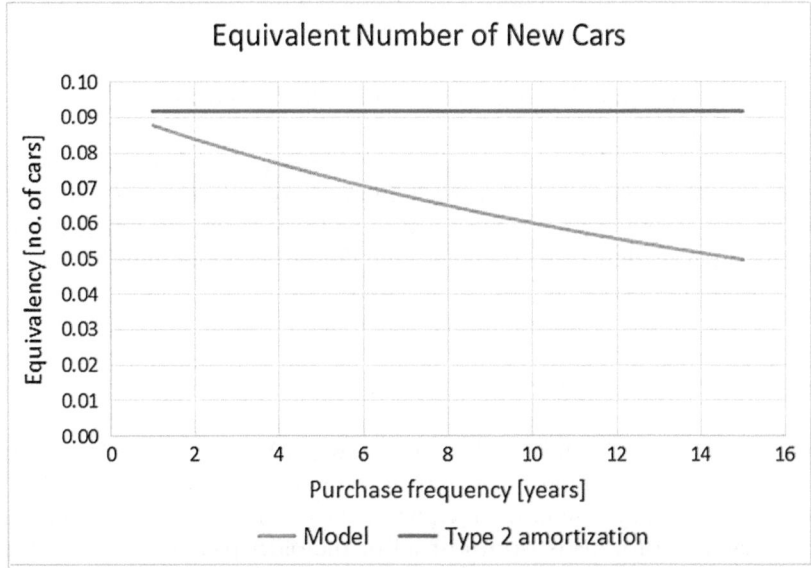

Fig. 3.2 Equivalent number of new cars as a function of new car purchase frequency. In Austria, based vehicle registrations, the average age of the automobile fleet is 7.4 years, and the average age of the fleet in the EU in 2018 is 10.9 years. The difference is due to the export of used cars. (Source: Lead author's unpublished self-calculation)

3.3 MELDING BOTTOM-UP FOOTPRINTS WITH TOP-DOWN AVERAGE PER PERSON EMISSIONS

As I have tried to show, there is no "right" or "wrong" method for GHG footprinting. The analyst has to make decisions on what activities and consumption items to include and how to include them. These decisions are based on the analyst's perception of the individual's knowledge and time-tolerance, prior understanding of the important components of the footprint in her/his country and wish to focus on a specific lifestyle item. For example, micro-plastics have recently been in the public focus. They may be an item in current and future footprint analyses but were probably not considered in historical studies. These decisions lead to inaccuracies in the estimated total footprint, Eq. 3.2, as compared to top-down average per person emissions. The reasons are many-fold, as follows:

1. There is truncation error (Eq. 3.2).
2. The efficiency indicators and emission factors (Eq. 3.1) may be an outdated or not suitable for use in the analyst's country.
3. The amount of service (Eq. 3.1) provided by the consumer may be inaccurate due, among others, to

   (a) Lack of knowledge of the consumer,
   (b) Lack of patience by the consumer,
   (c) Misinterpretation by the consumer of the analyst's data need, and/or
   (d) Discretization by the analyst of the answer choices for the consumer so as to improve consumer participation (i.e., reduce inaccuracies a and b above).

So, how does one correct for this lack of accuracy? As a general strategy, I recommend that the analyst honors as many top-down values as possible and uses the bottom-up footprint analysis to provide lifestyle-based "color." This means that, in addition to the footprint study, one should have a set of results from a representative sample of the population.

To correct the bottom-up results to top-down, one could add a constant or apply a scaling factor to the bottom-up amount of services, or both. A problem with shifting the amount of services by a constant is that some services have a "true" zero, and a shift does not respect this "true" value. For example, the top-down mean amount of beef consumed is

92.5 g/person/day and the mean value from the representative sample is 85 g/person/day due to any or all of the reasons listed above. Then, the vegetarians and vegans will be not very happy if they are assumed to eat 7 g/person/day of beef due to the melding process. For services that have a normal distribution, both the application of a shift and of scaling achieve the new mean value. However, scaling is a better correction when applied to data that have a log-normal distribution, and many services or products approximate this distribution. With only one value as a benchmark (the top-down mean), one cannot calculate a scale and a shift.

I recommend that the analyst correct as many service or consumption values as he/she can to a top-down benchmark (Eq. 3.1) and make a second check against emission benchmarks if possible. The second check would correct against differences in the values of the efficiency indicator and emission factor (Eq. 3.1).

3.4   An Actual Example

As an example, let me explain the method I used for footprinting diet in the EU H2020 project, ECHOES. In ECHOES (https://echoes-project.eu/), our goal was to estimate the personal energy demand via a survey for individuals in all EU countries. The entire detailed energy footprinting methodology is available in Bird et al. (2019). ECHOES food footprinting model includes 18 food types.[3] As is the case, our survey was time limited to 10 minutes. So, we had space for only one question on diet. We asked the following:

Q: Please choose the answer that best describes your diet.

1. Meat in most meals
2. Meat in some meals
3. Meat very rarely
4. No meat, but fish
5. Vegetarian
6. Vegan

---

[3] The 18 food types are: Beer, Cheese, Coffee, Eggs, Fish, Fruit, Grains, Meat_beef, Meat_chicken, Meat_pork, Milk, Nuts, Oil, Potatoes, Pulses, Sugar, Vegetables, and Wine. There was no distinction made between individual grains, vegetables, and fruits, as they have very similar LCA emission factors.

As my top-down control, I used personal food consumption statistics from the Food and Agriculture Organization (FAO) of the United Nations (2021). I chose these data because (1) I needed consistency across the entire EU, and (2) the FAO data can include post-production losses and amounts for human consumption. These two points are important because one needs to match the consumption data with the assumptions made in the LCA emission factors. The emission factors assumed no post-production losses and that animals consume feed. It is important to note that I did not assume nationally recommended consumption based on nutritional studies. Those are what we *should* eat, not what we *do* consume (eat + losses).

The methodology affects only the meat consumption and what is consumed in its place. For other food items, each respondent received the national average. To start, I assumed that vegans consumed the nutritionally recommended foods for a vegan diet. Then I assumed that answer 1 fit the national average meat consumption, that answer 5 (vegetarian) indicates eating no meat and that answers 2 and 3 indicate consuming 67% and 33% of the national average, respectively. For groups that ate less than the national average, the foregone calories provided by meat were replaced by nuts, oils, and pulses (spread equally across the three food groups). For the pescatarians (answer 4), I assumed that all meat calories were replaced by fish calories.

At this point, it should be clear to the reader that, with respect to the top-down control, I have underestimated the consumption of meat and overestimated the consumption nuts, oils and pulses. So, now I apply a scalar to the individual consumption of each food type so that the survey average value per food type equals the top-down control. It is important to note that using a scalar, the vegans and vegetarians still have no meat consumption. The answer 1 respondents, after correction, consume more meat than the national average. As a result, using a single question that is easily answerable by all respondents, I have created a realistic distribution of the consumption of food types. Even though the question has a subjective answer, the distribution has the correct average value. The adjusted consumption by food type was applied in Eq. 3.1 using EU-average LCA energy factors for each of the 18 food types. To make a GHG footprint, one need only use EU-average LCA emission factors instead of energy factors.

A point to consider is whether the top-down control should be a consumption value or an emission value. Here, the reader should remember that if she/he applies LCA emission factors in Eq. 3.1, some of the emissions might occur outside the national inventory. Balancing to the national inventory is only possible if the fraction of the emissions that occurs as part of the national inventory can be ascertained.

Now, I am certain that readers will be critical. You will undoubtedly suggest that more detail is needed, and I agree. However, the example illustrates the footprinter's conundrum. We were limited in regard to how many questions we could ask, and the survey respondent may not have good knowledge of the answer. For example, can you really tell me how many kilograms of wheat you consume in a year (including food waste)? In addition, in ECHOES, we were interested in the full personal energy footprint, and other segments of the economy, specifically consumption, housing, and transportation, were considered more important. Also, these are sectors where the respondent has good knowledge.

3.5   Conclusions and Outlook

In this chapter, I have presented my recommendations for calculating GHG emission footprints. I have tried to stress that there is really no "right" or "wrong" method to do this. The analyst must decide what type of emissions to use and which items and services to include in the footprint. He or she must clearly document these decisions so that the reader or user of the resulting footprint understands the limitations of the results. The decisions made by the analyst are made due to data availability, a priori understanding of the individual's response to the analyst's survey, consider the length of the survey and respondent's fatigue, and often include preselected focus areas for study. The next chapters discuss the development of a reliable questionnaire and the pitfalls associated with this approach. The derived questions and data were then used to estimate the GHG emissions and energy demand of the respondents based on the approach presented in this chapter. I have also tried to show that there are ways to correct the results so that they match top-down data and GHG emission estimates, which I believe should hold as benchmarks. Happy footprinting!

REFERENCES

Bird, D. N., Schwarzinger, S., Kortschak, D., Strohmaier, M., & Lettmayer, G. (2019). *A detailed methodology for the calculation of cumulative energy demand per survey respondent.* ECHOES project report 5.1.1. JOANNEUM RESEARCH Forschungsgesellschaft GmbH, Graz, Austria.

Čuček, L., Klemeš, J. J., & Kravanja, Z. (2012). A review of footprint analysis tools for monitoring impacts on sustainability. *Journal of Cleaner Production, 34*, 9–20.

Davis, S. J., & Caldeira, K. (2010). Consumption-based accounting of CO2 emissions. *PNAS, 107*(12), 5687–5692.

FAO. (2021). FAOSTAT: New food balances. Retrieved February 25, 2021, from http://www.fao.org/faostat/en/#data/FBS

Fuchs, R., Brown, C., & Rounsevell, M. (2020). Europe's green deal offshores environmental damage to other nations. *Nature, 586*(7831), 671–673.

IPCC. (2006). *2006 IPCC Guidelines for national greenhouse gas inventories.* Prepared by the National Greenhouse Gas Inventories Programme. Edited by Eggleston H.S., Buendia L., Miwa K., Ngara T., & Tanabe K. IGES, Japan.

Jungmeier, G., Bird, D. N., Lettmayer, G., Hingshamer, M., & Schwaiger, H. (2020). *Abschätzung der konsumbezogenen Treibhausgas-Emissionen der Stadt Wien: Status Quo, Lebensstile und Maßnahmen.* Report LIFE 2020/1. JOANNEUM RESEARCH Forschungsgesellschaft GmbH. Graz, Austria.

Liptow, C., Janssen, M., & Tillman, A.-M. (2018). Accounting for effects of carbon flows in LCA of biomass-based products—Exploration and evaluation of a selection of existing methods. *The International Journal of Life Cycle Assessment, 23*(11), 2110–2125.

Mulrow, J., Machaj, K., Deanes, J., & Derrible, S. (2019). The state of carbon footprint calculators: An evaluation of calculator design and user interaction features. *Sustainable Production and Consumption, 18*, 33–40.

Padgett, J. P., Steinemann, A., Clarke, J., & Vandenbergh, M. (2008). A comparison of carbon calculators. *Environmental Impact Assessment Review, 28*(2–3), 106–115.

Peters, G. P. (2008). From production-based to consumption-based national emission inventories. *Ecological Economics, 65*(1), 13–23.

Windsperger, A., Windsperger B., Bird D. N., Jungmeier G., Schwaiger H., Canella L., Frischknecht R., Nathani C., Guhsl, R., & A. Buchegger. (2017). *Life cycle based modelling of greenhouse gas emissions of Austrian consumption.*

Final Report of the Research Project ClimAconsum to the Austrian Climate and Energy Fund, Vienna. Institut für Industrielle Ökologie (IIÖ). St. Polten, Austria. https://www.klimafonds.gv.at/wp-content/uploads/sites/6/201 70913climAconsumACRP7EBB464796KR14AC7K11791.pdf

Wood, R., Neuhoff, K., Moran, D., Simas, M., Grubb, M., & Stadler, K. (2019). The structure, drivers and policy implications of the European carbon footprint. *Climate Policy, 20*(Sup 1), S39–S57.

CHAPTER 4

The Development of the Questionnaire

This chapter[1] presents the development of our survey questionnaire and the results of the subsequent validation efforts. The questions address the areas of climate-relevant behaviors discussed in the previous chapter and are based on the various existing scales, lifestyle calculators, and surveys, such as the ECHOES project introduced in the previous two chapters. We, however, go beyond these existing questions and questionnaires as we also include various new items and analyze the validity of existing and new questions. For this purpose, we conducted a survey in two waves and also collected additional material from our respondents. The final data set is available at the Austrian Social Science Data Archive (Hadler et al., 2021).

4.1   QUESTIONS INCLUDED IN OUR SURVEY

As pointed out in the previous chapters, we also aim to explain emission-relevant behaviors and thus also include questions on environmental attitudes and personal PEB and various socio-demographic variables since previous research has highlighted their significance (Diekmann & Preisendörfer, 1992, 1998; Stern, 2000; Dunlap & Jones, 2002; Gatersleben et al., 2002; Marquart-Pyatt, 2008; Huddart Kennedy et al., 2015). The general environmental behavior and attitude questions are mostly equivalent to those of the International Social Survey Programme (ISSP, 2019; www.issp.org) and the emission-specific items to those used

[1] Lead authors: Markus Schweighart and Rebecca Wardana.

© The Author(s) 2022   53
M. Hadler et al., *Surveying Climate-Relevant Behavior*,
https://doi.org/10.1007/978-3-030-85796-7_4

Table 4.1 Overview of areas within the questionnaire (first wave)

Housing	Building information	Direct
	Heating and heating behavior	energy
	Power consumption	
	Water treatment and water consumption	
Mobility	Individual motorized means of transport (car,	Direct
	motorcycle, and public transport)	energy
	Flight behavior	
Diet	Consumption of animal products	Indirect
	Waste	energy
Consumption	Consumption of goods (e.g., purchasing of clothing	Indirect
	and other goods)	energy
	Consumption of information (e.g., purchasing of	
	electronic devices)	
	Leisure activities	
Not emission-related	General environmental attitude	
items	PEBs	
	Socio-demographic as important influencing	
	variables	

in the ECHOES project (Reichl et al., 2019). Table 4.1 provides an overview of the areas for the measurement of individual emission-relevant behavior and other areas that are included in our initial survey. The six areas—housing, mobility, diet, consumption of goods, consumption of information, and leisure activities—are based on the considerations presented in Chap. 3. The detailed list of questions can be found in the Appendix of this book.

The follow-up survey focused mainly on the car use of the respondents since the time of the first survey. In the first survey, respondents were asked to indicate the mileage of their most frequently used car and for an estimate of how often they used it (both in kilometers and in hours). In the second wave of the survey, they were asked to estimate their mileage since the last survey and to provide their current mileage of the same car. This enabled an approximate projection of car use over the entire year, and these mileage figures could also be used to validate the respondent's own assessment. The following questions were included in the second wave.

- Do you live in ownership or rent? (refers to residential property at the first survey date)
- How many kilometers have you traveled since the last survey at the end of February/beginning of March? (estimate)

- What is the current mileage of your most used car? If this is no longer the car we asked you about last time, we do not need the mileage. (mileage at second time)
- How many times has this car been used by others since February/ March without you in it? (never, almost never [about 10% of the km], rarely [about 25% of the km], about half of the time [about 50% of the km], and often [75% of the km or more])

4.2   SAMPLES

The survey was conducted in two waves and focused on urban, suburban, and peripheral regions. Our sample (Hadler et al., 2021) includes the Austrian capital, Vienna, and the capital of the province of Styria, Graz, which is the second largest city of Austria, with about 300,000 inhabitants. Within these two cities, a new subdivision into "bourgeois districts" and "workers' districts" was made. One "bourgeois district" and two "workers' districts" were selected for the survey since it was assumed that the bourgeois districts would be more willing to participate. The areas around Vienna and Graz were selected for the suburban area. These regions were again subdivided according to size and accessibility in order to achieve a higher comparability, on the one hand, and, on the other hand, to provide good public transport connections for the interviewers. In order to also cover rural areas within the sample, more remote areas were sought. The two municipalities of Murau in Styria and Waidhofen an der Thaya in Lower Austria as well neighboring villages were chosen. Once the locations for the sampling had been determined, the respondents were chosen randomly from the online telephone book "Herold."

In the first survey wave, a total of 209 persons were interviewed in February and March 2019. Due to the complexity of the questions and the additional validation questions, the questionnaire in the first survey was filled out in face-to-face interviews. In total, there were 12 interviewers who were trained to administer the survey. The questionnaire was presented to the respondents by the interviewers, who had an extended version of the questionnaire that included the validation questions. The final answers were completed by the interviewers themselves. In addition, some of the respondents were asked how confident they were in answering the questions.

It was also specified that the first survey should be conducted in the respondents' private households as some questions required proof of certain receipts (e.g., heating and electricity bills) and included questions about the household's equipment (e.g., electrical appliances or insulation measures). If the respondents did not agree with taking the survey in their own household, a neutral location was suggested to them. In the end, only those persons who indicated the mileage of their most frequently driven car were asked whether they would be willing to be available again for a follow-up survey. The main objective of this second survey wave was to determine the mileage of the respondents since the first survey date.

Our sample for the first wave consists of 52.2% male and 47.8% female respondents. Overall, 36% of the sample come from urban areas (13.2% from Graz and 22.8% from Vienna), 41.6% live in the suburbs around Vienna and Graz, and 22.4% live in the countryside. The respondents are between 20 and 94 years old, with the average age being 55 years. The majority of the sample consists of predominantly older respondents. The distribution of the educational qualifications shows that 38.5% have a university degree, 27.8% have a high school degree, and 17.3% an apprenticeship certificate. Regarding income, both the individual monthly net income and the total monthly net income of a household were asked. The average net income of a person within the sample lies between €1751 and 2000 per month. The average total net income of a household is between €2751 and 3000 per month. Looking at the composition of households, 78.5% of respondents live without children under 18 years of age. Of these, 23.4% are one-person households, and 62.7% live in a two-person household (two adults). Only around 10% live with one or two children under the age of 18 in a household.

Comparing these figures with the socio-demographic distributions of Austria, it becomes clear that the sample collected shows an above-average representation of the older generation and people with an academic degree as well as an under-representation of people with an apprenticeship certificate. By comparison, the average age in Austria is 42.9 years, and the proportion of persons over 60 years of age is 25.4% (Statistik Austria, 2020a). In comparison, the share of people aged over 60 in the present sample is over 40%. Moreover, the average individual monthly net income of the sample is below Austria's average net income (€2226 per month; Statistik Austria, 2020b). Another over- and under-representation is also evident in the distribution of educational attainment. Austria-wide data show that a total of 17.5% of the Austrian population has a university degree and that apprenticeship qualifications are the most common educational qualification in Austria, with 34.1% (Statistik Austria, 2020c).

A total of 141 persons were eligible for the second survey wave on the basis of the first survey wave as they indicated that they use a car, and the second survey focused on this. In the end, a total of 68 persons were willing to take part a second time. They were contacted by telephone in October and November 2019. They were given the opportunity to conduct the interview by phone or online via a link if they did not have time to answer at the time of the call. They were asked a total of five questions focusing on individual car use since the last survey. At the end of the survey phase, all respondents were sent an individual CO_2 profile, if desired.

A comparison between the socio-demographic distribution of the first and second survey waves shows that more men than women were reached in the second survey wave. In addition, significantly more older people were included in the second survey wave. Persons in suburban and rural areas were reached more often in the follow-up survey than persons in the city. The distribution of educational attainment is similar in both survey waves although it should be noted that in the second survey wave, significantly more persons with a vocational higher education degree (BHS) were reached compared to those with a general higher education degree (AHS).

4.3   VALIDATION OF OUR QUESTIONS

A central issue of the project concerns the validation of questions regarding the suitability for the collection of GHG emissions. In the present case, a question is valid if it refers to behavior that is associated with GHG emissions and if it actually measures this behavior. The first point refers to the content relevance of the questions with regard to the GHG emissions caused by the respondent. Chapters 2 and 3 dealt with this *content validity* by reviewing the literature on the topic of GHG-relevant behavior and presenting the empirical findings to date as well as through calculations using the LCA approach. The areas that are most significant in terms of emissions have been identified, which are mobility, housing, diet, and consumption. It is only for these areas that it makes sense to formulate questions to ascertain emission-relevant behavior. In order to verify that the questions actually measure the named behavior, different question variants can be compared with each other and an external criterion. In this way, criterion validity is assessed. Additionally, participants can be asked directly whether or not they think that their responses are accurate to also cover the subjectively assessed validity.

The following sections now address these validity issues for the various relevant behavioral domains. The procedure thereby will always include the same points, as follows: (1) a comparison of various question variants regarding answer distributions, missing answers, and so on; (2) comparison with an external criterion; and (3) the results of the respondent's self-assessment of how confident they were with the answers they gave (where possible[2]). For this last aspect, respondents were asked after the regular survey how confident they were with their answers to certain questions on a 5-point response scale ranging from "very confident" to "very uncertain."

Housing

To determine the GHG emissions due to space heating, a number of variables were collected and used for the calculation, including the living space, the number of people living in the household, the type of dwelling (single-family house and similar), the age of the dwelling, the degree of insulation, the main energy source used, the type of heating system, and various behavioral variables, such as turning the heating down in different cases or the temperature at which the heating is applied. In addition, the heating costs per month and the heating energy consumption, which is listed in the heating bill, were also asked. In addition, the temperature in the living room was measured by thermometer during the survey.

The main influencing factors regarding heating emissions per capita are the floor space, the number of persons living in a household, and the type of main source of energy used (thus the kind of fuel) (Schweighart et al., 2020). If the type of dwelling and the thermal insulation are added, the explanatory power for heating emissions increases to approximately 70%. This value can only be increased insignificantly by adding behavioral variables, such as turning down the heating or temperature control. This does not mean, however, that the influence of the individual is not present in this area. Rather, the possibility of influencing their heating-related footprint can be found in the decision regarding their own living situation. Since a subsequent thermal refurbishment or a change of the main energy source is often not possible without high costs, the choice of apartment is of great importance. Since, however, ecological motives are unlikely to be the decisive factor in the decision for or against a particular form of

---

[2] Due to the already demanding survey format, we refrained from asking respondents for this self-assessment for each question.

housing, but rather financial, occupational, and lifestyle reasons, the effect of ecological attitudes on actual emissions can nevertheless be classified as comparatively low in this area.

Heating emissions were estimated based on these technical variables. To validate these numbers, we used the heating energy demand and heating costs found on energy bills. The results show strong correlations with the calculated magnitude of heating emissions even though the validation criteria do not include the main energy source as an essential factor. The correlation between heating costs and heating emissions is stronger ($r = 0.67$, $p < 0.01$) than the correlation between the energy demand according to the heating bill and the heating emissions ($r = 0.53$, $p < 0.01$). However, the number of those who could or wanted to present a heating bill was relatively small (29% or 14% of all respondents), which is why more detailed analyses were not carried out here. However, the strong correlation with the calculated quantity speaks for the validity of the calculated quantity and the central variables used. Regarding the monthly heating costs, more data are available as 83% of the respondents provided this information. Therefore, more detailed models could be calculated here. The floor space, the number of persons in the household, the thermal insulation, and the dwelling type turned out to be the strongest predictors of heating costs per capita. Respondents also expressed a high degree of confidence in the accuracy of their statements. For example, 82% are "very confident" about the accuracy of the living space they reported. In summary, the results are a clear empirical indication of the relevance and validity of those questions.

To cover not just technical features but also a behavioral aspect of housing, we will now deal with room temperature setting. This was included in the questionnaire in two different ways—by asking (1) what temperature the most frequently used room is heated to during the day and by asking (2) how the apartment temperature is assessed compared to other apartments (5-point scale from "considerably lower temperatures" to "considerably higher temperatures").

Additionally, the interviewers placed a thermometer in the living room during the interview.[3] The results show that the self-reported and the measured temperature are strongly related ($r = 0.58$, $p < 0.01$). For explaining heating costs, the thermometer measurement exhibits a clearly higher portion of explained variance than the self-reported temperature.

---

[3] The survey period in February and March 2019 falls within the heating period.

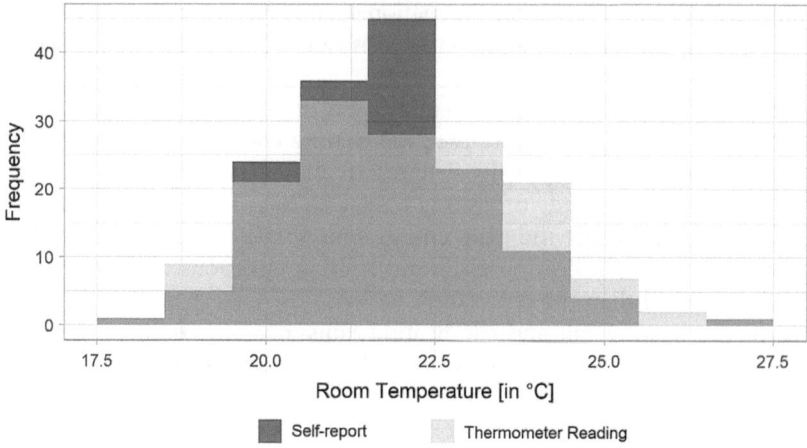

Fig. 4.1 Distribution of temperatures according to self-reported data and thermometer measurement. (Source: OeNB sample Hadler et al., 2021)

The punctual measurement by thermometer could therefore supply better results because here neither social desirability nor fuzziness in communication (e.g., in understanding the question) plays a role.

Comparing the distributions of the self-reported and measured temperatures (Fig. 4.1), it can be seen that more temperatures above 22.5 °C were recorded by thermometer and that the self-declarations are overrepresented in the range between 19.5 °C and 22 °C. In addition to a simple underestimation of the temperature, social desirability could also be an explanation in that one wants to appear to be energy-saving. Another explanation for this could be that the thermometer measurement was only taken at a certain point in time and that the presence of several people in the living room also leads to a slightly higher temperature.

Regarding the question about the temperature estimation compared to other apartments, we find that the answer category "considerably lower temperature" was not chosen by the respondents at all and that approximately 50% put themselves in the category "as average." If we compare the measured temperatures given for each category, we can see that people who state that they have warmer homes than others actually heat to higher average temperatures but that the difference between the categories "lower than average" and "average" is very small. This question therefore allows for fewer meaningful distinctions.

Interestingly, the argument that this question has fewer categories but is easier to answer does not apply as fewer respondents say they are "very confident" about the accuracy of the information they provide compared with the question on the exact temperature (43% vs. 58%). In short, the question on the assessment of the temperature compared to other homes allows fewer meaningful distinctions as it is less suitable from this point of view for depicting the interior temperature as an aspect relevant to heating energy.

On the basis of these findings, the questions about the living space, the main energy source, and the thermal insulation prove to be well suited to approximate heating emissions since they assert the largest influences. Regarding behavior, the temperature setting appears to have the largest effect and is best determined by asking directly about the temperature.

Mobility

The *intensity of car use* was surveyed with two variants—one question asked about the kilometers traveled in the previous year, and another asked about the average time spent in the car per week. The phrasing of the questions themselves shows that somewhat different things are addressed. The question about kilometers traveled clearly contains a time frame in the question, namely the previous year, and aims at an estimation of the total distance in kilometers. This means that non-routine journeys, such as those made during vacation periods, are also included. The question about the time per week in the car does not specify a clear time period, so that reference is more likely to be made to current usage. The addition of "on average" opens a certain space for interpretation—the person questioned is free to choose the period over which he or she calculates the average. Furthermore, "on average" can be interpreted as an indication to make a rough estimate. Also, the interpretation that the question refers to an average week is possible, whereby then again vacation trips would not be covered.

An indication that something different is being measured with the two questions can be found in different response patterns. To get an idea of this, the empirical answer distributions are shown in Fig. 4.2.

The main difference, which becomes apparent from the comparison of the answers, concerns the shape of the distributions. While the question about annual kilometers leads to a single peak distribution, which is scattered around the most common category (5001–10,000 km), the

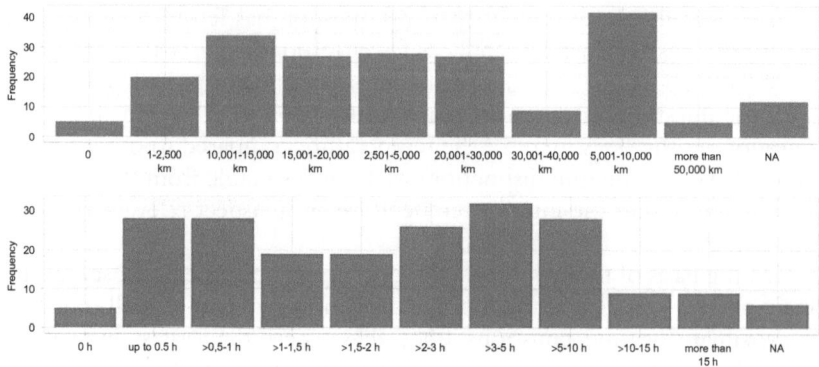

Fig. 4.2 Response distributions (absolute frequencies) for estimation of kilometers driven or hours in a car. (Source: OeNB sample Hadler et al., 2021)

question about hours per week shows two peaks, one in the range between 0.5 and 1 hour and one at 3–5 hours. Both questions seem to have ample answer options in the upper-scale range—that is, for people who travel comparatively often by car. However, in the lower-scale range—for those who use their car infrequently—the question about weekly hours has a second answer category—"up to .5 hours"—that is rather broad. In the very low answer spectrum, this question is therefore less able to differentiate. In part, the differences between the answer distributions are also due to the fact that the distances between the categories are different. For the questions on hours spent, the category width shows a larger spectrum—from 0.5 hours with the second category ("up to .5h") to 5 hours with the tenth category ("> 10–15 h").

However, the two distributions also show similarities. In both cases, five people state that they do not use cars at all. The number of refusals (or "don't know") is slightly higher when asked about kilometers traveled (6% compared to 3%). This is interesting because, initially, it was assumed that the question about the kilometers traveled during the year might be difficult or even impossible to answer for some people. Even though it has been shown that more people do not answer the question about yearly kilometers (12) than about weekly time in the car (6), the level of answers that cannot be used for calculations remains relatively low.

For criteria validation of the car-related questions, the difference between the mileages at both survey points was used as a criterion. At both points in time (Feb–Mar 2019 and Oct–Dec 2019), the kilometer reading of the most frequently used car was asked. Since it is furthermore possible that a car is also used by others, the second time the survey was conducted, the question was asked as to how often (how many % of the total km) someone else used the car without the respondent him/herself being on board. Then, these answers were used as a weighting factor for the mileage difference between the two points in time to ensure that the result approximates the driven distance when the interviewee was also in the car as a driver or passenger, which is what was asked by the questions to be validated beforehand. So, the question about annual kilometers is refused slightly more often, but it differentiates answers somewhat better, and respondents are more confident in their answers, which is why it seems to work better in surveys in this context.

To check which of the two questions on car use intensity is better suited to approximate the actual behavior, we compared them with an external criterion in the form of the mileage reading difference we had obtained for two dates. Pearson correlation analyses show that there is a strong correlation between the annual kilometer estimate and the kilometer difference ($r = 0.52$, $p < 0.01$). The correlation between the hours per week and the mileage difference is weak and not significant ($r = 0.25$, $p = 0.119$).

Finally, the interviewees were asked how confident they were in their answers for both questions. The comparison of this assessment shows that, on average, people think they are more confident when asked about annual kilometers traveled—63% say they are "very confident" about the accuracy of their answer, while only 44% are confident about the hours per week. Against this background, the assumption that people who are incapable of estimating car kilometers will nonetheless give some kind of answer for reasons of social desirability seems unfounded.

In summary, the question "How many kilometers have you covered in the last 12 months with a car? (as a driver and/or passenger)" can be considered valid according to the information available. It proves to be superior to the question on the time spent per week in a car in three respects. Firstly, the respondents are on average more confident of their answers than with the other question. Secondly, the connection between this question and the validation criterion is stronger than for the hourly question. Finally, considering the content, an argument can be made for the kilometer question because it covers non-routine car usage (e.g., on vacation), which the other question does not.

Alongside car travel, we also surveyed *personal air travel* in several variants. In one variant, the absolute number of flights in the previous year was asked about, distinguishing between different purposes (private and business flights) and distances (short distances up to 3000 km and long distances over 3000 km). Another variant asked about hours spent on an airplane during the previous year. Ten response categories were used, ranging from "0 h" to "more than 50h." As a last variation, there was a question asking which of six statements best describes one's own flight behavior.

Looking at the response distributions, it must be noted that over 50% of all respondents had not taken any flights at all in the previous year. When asked about flight time, many responses were concentrated in the "2.5 h–5 h" category, which roughly corresponds to the flight time of a round trip for a typical summer vacation. The distribution of responses to the question about which statement best describes one's own flying behavior shows that apart from the answer with the highest volume of flights, all other five response categories were chosen by 10–25% of respondents, with the top category being "I fly abroad once every few years."

To get a basis for the validation criterion, the respondents were additionally asked after the regular survey to name the origins and destinations of all flights they had taken in the previous year. Building on that, the average flight times were determined via Google and added. Finally, based on these annual flight times, flight-related CO_2 equivalents, which now serve as a validation indicator, were calculated analogous to the GHG emission calculation described in Chap. 3.

To determine which of the mentioned question variants best approximates flight-based GHG emissions, separate linear regression models[4] (ordinary least squares [OLS]) were calculated and their variance explanations compared. If only the total number of flights is considered, about 67% of the variance was explained (adjusted R^2s). Since intercontinental flights produce significantly more emissions than short-haul flights, it makes sense to make a corresponding distinction. This step increases the proportion of explained variance to 81%. A similar amount of variance, namely 78%, can be explained when the question on the annual flight time

---

[4] When asked about the number of flights, simply the numeric values (for short- and long-haul flights) were used as predictors. For the categorical question about flight time, the category midpoints were used as numerical values. For the statements, dummies were calculated for five of these statements, and "I never fly" was used as the reference category.

is used. The advantage of this is that only one question is necessary, but it is still possible to somehow differentiate by flight length, which has a strong and direct[5] influence on GHG emissions. However, this question is cognitively rather difficult to answer because, theoretically, all individual flight times must be estimated and added.

Interestingly, the question about individual flight behavior with only six statements as answer categories also explains a quite high proportion of flight-based emissions (48%). The regression analysis shows that the statements ranked according to increasing flight volumes are also associated with correspondingly higher regression coefficients. Compared to the reference category of non-flyers, especially the categories "1 time per year," "several times per year (short distance)," and "several times per year (also long distance)" exert significant influences on the calculated flight-based emissions.

However, the direct comparison shows that this question is significantly worse at approximating GHG impacts than those questions based on a numerical assessment of flight behavior (number of flights or hours). When asked how confident respondents are about their given answers, 88% said they were "very confident" when asked about the number of flights, whereas only 79% of respondents gave this answer when asked about flight hours.

The validity assessment in this section concludes that the two questions about the number of short- and long-distance flights allow the best approximation to the validation criterion. Since these also seem to be somewhat easier to answer than the question about hours in airplanes, they have proven to be the most appropriate here.

Other forms of mobility, such as the use of public transport, bicycle riding, or walking, were not considered as these involve only very small amounts of GHG emissions per person compared with the use of cars or airplanes.

---

[5] The effect of flight time on flight-based GHG emissions is, of course, mediated by the size of the aircraft, the number of passengers, and the energy efficiency of the engine. However, since these are far from being collected by the questionnaire, they must be left out here.

Diet

Regarding food-based behavior, the interviewees were asked a general question about their meat consumption with answer categories (ranging from "meat in most meals" to "no meat at all"), a question about the frequency of restaurant visits and the like, and a question about what proportion of food they throw away on average. As a validation criterion, the consumption frequency of energy- and resource-intensive foods was surveyed in detail. For example, the frequency of the consumption of different types of meat, fish, cheese, and eggs was asked. Based on this information, CO_2 equivalents were calculated for each person.

It can be seen that those who state that they eat meat only very rarely or live vegetarian or vegan lifestyles have significantly lower nutritional emission values. The difference between those who eat meat in some meals and those who eat meat in most meals is less pronounced. However, it can be seen that it is above all the consumption of sausage and pork in general that decreases the most in people who state that they consume less meat. The consumption of fish, but also beef, on the other hand, decreases less. This is also interesting because it is mainly beef that is associated with particularly high GHG emissions per kg of meat. It seems to be that less exclusive types of meat, such as sausage and pork, are being avoided.

To evaluate how strongly the variable on eating habits affects food-based emissions, an OLS model was calculated, revealing that the answer categories (with "meat in most meals" as the reference category) explain about 32% of the variance (adj. R^2) of food-based emissions. The question about eating habits thus seems to be a useful indicator for the GHG consequences of individual nutrition. However, it was also shown that it is the consumption of special animal foods that is particularly effective here. As an alternative to the question about eating habits, a direct question about the frequency of the consumption of different foods can be used.

The frequency of restaurant visits also shows a certain effect—those who eat out more often have significantly higher emission values because they consume emission-intensive food more often. However, this question can only explain 7% of the variance in food-based emissions, which is why this question is less relevant when it comes to collecting data on GHG emissions.

Consumption

This area includes questions about shopping behavior regarding goods for personal use other than food, which has already been discussed. Here, we deal with the consumption of goods but also with the use of electronics and with leisure activities. The production and transport of such consumer goods generate GHG emissions, which would have to be recorded individually and added up to obtain the individual emission values in this category. However, there are a many sub-categories (e.g., for clothing: shoes, pants, coats, socks, etc.) that are difficult and costly to survey.

With regard to electronics, the respondents were asked to choose one of five statements that most closely correspond to their purchasing behavior of electronic items, ranging from "I don't need most of it" to "I make sure I always have the latest technology." The clothing-related question asked about the respondents' approach to clothing (5 categories, from "very modest" to "always in the latest style").

Since there is no clear external validation criterion here, similar to the diet area, the approach taken was again to use a detailed survey of the consumption frequency of certain important goods. To do this, we surveyed the frequency of purchases of smartphones, laptops, televisions, and so on and asked respondents about the number of shoes, pants, coats, and other clothing categories they purchased in the past year. This information was then in turn used to calculate corresponding GHG emissions.

Looking at the frequency distributions for the electronics variable, it is noticeable that no one chose the extreme category "I make sure I always have the latest technology" and that there is little variation in the responses, with three out of four respondents choosing "I take care to use it for a long time and replace electrical items only when they break." When asked about the use of clothing, there was also no one who answered, "always in the latest style." Here, however, the answers are more strongly distributed among the other categories. Interestingly, the most frequent choice was "long use" and not the always-appealing answer in surveys "average."

With regard to electronics articles, the comparison with the calculated emission quantity (validation criterion) shows that those who say that they "don't need" most electronics products are hardly any different from those who "pay attention to long-term use" and "buy new equipment from time to time." Only those who state that they "regularly" buy new electronics come up with significantly higher emission values. Despite five response categories, this question empirically distinguishes basically just two levels when it comes to the related emissions (as mentioned, one category was not selected at all). Again, socially desirable response behavior could play a role here. Especially vaguely formulated answer categories allow to choose supposedly "desired" answer categories based on favorable interpretations. This question is ultimately not well suited to survey emissions-related consumption aspects.

While it turns out that the response categories for the question on clothing are relatively well suited to reflect the intensity of the actual purchasing behavior of clothing, the influence on the total GHG emissions from clothing is not particularly strong. This means that there is no strong correlation between personal emissions attributable to clothing purchases and the clothing questions with the five statements.

To get an impression of which of the questions dealt with is best suited to explain the goods consumption-based emissions, regression models were calculated additionally. These models show that the question on clothing has the most explanatory power for this (8% adj. R^2), followed by the question on the use of electronics (7% adj. R^2). However, what also emerges from these calculations is that it is the consumption of "cars" that makes an even more central contribution to explaining these emissions[6] given the impact from the production of an average passenger car, which is estimated to be about seven tons of CO_2 for a small passenger car (Kawamoto et al., 2019). With an average use of a car of ten years (about 40% of the people in the sample say that they buy a car less often than all of ten years), this is still 0.7 tons of CO_2 per year, which can be attributed solely to the production of the car. Accordingly, it is not surprising that the question of how often a new car is purchased contributes even more to explaining the emissions caused by the consumption of goods than those mentioned so far. However, because it is difficult to ask questions about infrequent, large purchases, and because there are also many factors involved in the question about the inventory of vehicles that are difficult to take into account (age of the car at the time of purchase or shared cars), such questions remain difficult in surveys as a source for estimating individual environmental impact.

---

[6] The share of the explained variance for this variable in the emissions attributable to the consumption of goods is 40% (adj. R^2). Yet, a direct comparability with the other variables mentioned is not given since the purchase frequency of a car is directly included in the calculation of the validation criterion.

4.4   Conclusions and Outlook

This chapter described our questionnaires and samples in detail and presented the results of our validation efforts. Based on various aspects of this validation, we recommend the use of the items on living space, main energy source, thermal insulation, and temperature setting in the area of housing. As for mobility, we recommend asking about the distance traveled by car in the previous year to capture the intensity of car use and to ask separately about the number of short-haul and long-haul flights taken in the previous year to capture personal air travel. In regard to diet, either the use of a general question about dietary habits with formulated statements as response categories or a question about the frequency of the consumption of particularly energy-intensive foods is recommended. For consumption, a question about the purchasing behavior of new clothing can be used to distinguish the largest consumption levels that are relevant in terms of GHG emissions.

So far, we have considered all items and areas separately. The following chapter will look into the question of how to explain the total GHG emissions of our respondents. Firstly, we consider which items are the most suitable for this purpose, with the aim to find a highly parsimonious model. After all, it is not always feasible to include dozens of items on GHG emissions in a single survey, especially when environmental attitudes, sociodemographics, and other items need to be included as well. Secondly, the following chapter will also look into the question as to which factors have the strongest impact on the total GHG emissions of a respondent.

References

Diekmann, A., & Preisendörfer, P. (1992). Persönliches Umweltverhalten: Diskrepanzen zwischen Anspruch und Wirklichkeit. *Kölner Zeitschrift für Soziologie und Sozialpsychologie, 44*(2), 226–251.

Diekmann, A., & Preisendörfer, P. (1998). Umweltbewußtsein und Umweltverhalten in Low-und High-Cost-Situationen: Eine empirische Überprüfung der Low-Cost-Hypothese. *Zeitschrift für Soziologie, 27*(6), 438–453.

Dunlap, R. E., & Jones, R. E. (2002). Environmental concern: Conceptual and measurement issues. *Handbook of Environmental Sociology, 3*(6), 482–524.

Gatersleben, B., Steg, L., & Vlek, C. (2002). Measurement and determinants of environmentally significant consumer behavior. *Environment and Behavior, 34*(3), 335–362.

Hadler, M., Schweighart M., & Wardana, R. (2021). *OeNB CO2-relevant environmental behavior.* Data will be available for free at the Austrian Social Science Data Archive (www.aussda.at). https://doi.org/10.11587/WQGMKY

Huddart Kennedy, E., Krahn, H., & Krogman, N. T. (2015). Are we counting what counts? A closer look at environmental concern, pro-environmental behaviour, and carbon footprint. *Local Environment, 20*(2), 220–236.

ISSP Research Group. (2019). *International Social Survey Programme: Environment III—ISSP 2010.* GESIS Data Archive, Cologne. ZA5500 Data file Version 3.0.0. https://doi.org/10.4232/1.13271

Kawamoto, R., Mochizuki, H., Moriguchi, Y., Nakano, T., Motohashi, M., Sakai, Y., & Inaba, A. (2019). Estimation of CO2 emissions of internal combustion engine vehicle and battery electric vehicle using LCA. *Sustainability, 11*(9), 2690.

Marquart-Pyatt, S. T. (2008). Are there similar sources of environmental concern? Comparing industrialized countries. *Social Science Quarterly, 89*(5), 1312–1335.

Reichl, J., Cohen, J., Kollmann, A., Azarova, V., Klöckner, C., Royrvik, J., Vesely, S., Carrus, G., Panno, A., Tiberio, L., Fritsche, I., Masson, T., Chokrai, P., Lettmayer, G., Schwarzinger, S., & Bird, N. (2019). *International survey of the ECHOES project* [dataset]. Zenodo. https://doi.org/10.5281/zenodo.3524917

Schweighart, M., Schwarzinger, S., & Bird, D. N. (2020). Estimating heating-related GHG emissions: The advantage of a household composition-based survey approach. *International Journal of Sociology, 50*(6), 473–494.

Statistik Austria. (2020a). *Demographische Abhängigkeitsquotienten und Durchschnittsalter seit 1869.* Retrieved June 7, 2020, from https://www.statistik.at/web_de/statistiken/menschen_und_gesellschaft/bevoelkerung/bevoelkerungsstruktur/bevoelkerung_nach_alter_geschlecht/index.html

Statistik Austria. (2020b). *Nettomonatseinkommen unselbständig Erwerbstätiger nach sozioökonomischen Merkmalen—Jahresdurchschnitt 2018.* Retrieved December 16, 2019, from https://www.statistik.at/web_de/statistiken/menschen_und_gesellschaft/soziales/personen-einkommen/nettomonatseinkommen/index.html

Statistik Austria. (2020c). *Bildungsstand der Bevölkerung im Alter von 25 bis 64 Jahren, 1971 bis 2017.* Retrieved June 28, 2019, from https://www.statistik.at/web_de/statistiken/menschen_und_gesellschaft/bildung/bildungsstand_der_bevoelkerung/index.html

Stern, P. C. (2000). New environmental theories: Toward a coherent theory of environmentally significant behavior. *Journal of Social Issues, 56*(3), 407–424.

CHAPTER 5

Estimating and Explaining the Greenhouse Gas Emissions

This chapter[1] starts with an overview of the national emission figures for Austria according to the latest climate protection report at the time of our survey (Klimaschutzbericht, 2020). Subsequently, the emissions of our respondents are presented in detail and also contrasted with the number of national emissions. With a view on improving survey research, a validity check is performed to select those variables that allow an estimation of the total emissions caused by a person. The goal is to select the smallest number of questions that allow for a good estimate of overall emissions. Finally, this chapter seeks to explain the caused emissions from a social science perspective by using socio-demographic variables and environmental attitudes and behaviors as explanatory variables in a linear regression model. It concludes with a brief discussion of possibilities to reduce emissions.

5.1   Greenhouse Gas Emissions: The Case of Austria

In Chap. 3, an overview of how to identify emission-relevant areas and how to calculate them was given. Following the distinction between a nation's greenhouse gas (GHG) inventory and emissions produced through supply chain activities, this chapter seeks to present the latest amount of emissions produced in Austria at the time of our survey.

---

[1] Lead authors: Rebecca Wardana and Markus Schweighart.

© The Author(s) 2022
M. Hadler et al., *Surveying Climate-Relevant Behavior*,
https://doi.org/10.1007/978-3-030-85796-7_5

According to the climate protection report (Umweltbundesamt, 2020), Austria emitted a total amount of 79 million tons of CO_2 equivalents in 2018, when air traffic is included. The figure, which does not consider the European Emissions Trading System (ETS), according to the Climate Protection Act, sums up to 50.5 million tons of CO_2 equivalents. Compared to the year 2017, in which 82.3 million tons of CO_2 equivalents (including ETS) were emitted,[2] emissions had decreased slightly. This results in per-capita GHG emissions of 9.4 tons[3] (Statista, 2020a).

The climate protection report presents the share of individual sectors in total emissions for the year 2018 and therefore only refers to the national GHG inventory. The sectors with the highest share of emissions are energy and industry[4] (43.4%), transport (30.3%), buildings (10%), and agriculture (10.3%). A total of 3.2% of emissions are from waste management and 2.9% from fluorinated gases (Umweltbundesamt, 2020).

How does the individual behaviors of Austrians weigh up to the millions of tons of GHG emissions produced? Considering ETS, the transport and building sectors are responsible for large shares of the emissions caused. The individual behavior of Austrians is particularly noticeable in these two sectors.

The transport sector includes road traffic (divided into passenger and cargo transport), rail, maritime, national air transport, and military vehicles. The main emitter is road traffic, which accounts for 97% of the total emissions of the transport sector. Within road traffic, 63% of the caused emissions are produced by passenger transport, which is 18.8% of the total national emissions. Passenger transport includes the individual use of motorized vehicles (cars, motorcycles, and mopeds), public transport (bus, train, and tram), national air transport, and non-motorized means of transport (walking and cycling). Most of the emissions from passenger transport are attributed to car use (Umweltbundesamt, 2020). The individual car use of Austrians is thus responsible for a considerable part of the emissions from the transport sector.

In the buildings sector, emissions from the burning of fossil fuels for heating and hot water in private households and public and private

---

[2] 51.7 million tons without ETS.

[3] GHG emissions per capita according to the Climate Protection Act amount to 5.9 tons.

[4] Including those companies that are part of the ETS. If these were not taken into account, the total share of the energy and industry sector would amount to only 11.6%; transport (47.3%) is the largest emitter, followed by agriculture (16.2%) and buildings (15.6%).

services are added. Electricity consumption within the household is also considered. At the national level, this represents a share of 10%. Here, too, the main sources of these emissions are private households. In full, 83.3% of the emissions in this sector come from private households, which is 8.3% of the total national emissions (Umweltbundesamt, 2020).

In addition to central questions on living and mobility behavior, our conducted survey also included detailed questions on the respondents' consumer behavior, with a focus on electronics, clothing and leisure behavior. It must be assumed that many daily consumer goods are not included in the national GHG inventory if they have a production chain that extends beyond the national area. This is where consumption-based approaches (CBAs) and life-cycle assessment (LCA) come into play (see more details in Chap. 3), which attempt to calculate the GHG emissions generated by the consumption of goods within a country using the international supply chain and various input-output (IO) models. The latest research from Windsperger et al. (2017, 2019) has calculated that the consumption of goods added an additional 40% to the national inventory, referring to data from 2013.

In summary, these Austria-wide emissions data show that Austrians contribute a significant share to the total emissions with their individual behavior in the transport and buildings sectors. If the shares are added up, just over a quarter (27.1%) of Austria's total emissions are caused by the use of cars and residential buildings. As the conducted survey has also shown, mobility behavior (especially car use), as well as key figures for buildings, heating, and electricity consumption, provide basic data for calculating individual CO_2 consumption. Therefore, it makes sense to implement climate policy measures in these two sectors that lead to a rethinking or a transformation of previous behavior. With regard to consumer behavior, it is difficult to assign precise values on an individual level. Nevertheless, the calculations of Windsperger et al. (2017, 2019) show that emission savings in the global supply chain should also not be forgotten. From the national level of emitted CO_2, we are now coming back to the emissions of our sample. First, an overview of the sample's emissions is given.

5.2   Overview of the Sample's GHG Emissions

In total, an average of 7.36 tons of CO_2 equivalents are produced per person as a result of the behavior that was surveyed in our OeNB study (Hadler et al., 2021). As pointed out in Chap. 3, these calculations only include the emissions of activities and decisions in which the influence of the individual is predominant. They are directly caused by the individual's own behavior and do not include emissions from public infrastructure (e.g., roads), emissions from public services (e.g., administration and health care) and aliquot construction costs for residential buildings. These factors would result in an additional six tons of CO_2 per capita in Austria.

Table 5.1 presents the average distribution of the individual sectors in the composition of the total emissions based on our sample. The table shows that especially areas such as car use, meat consumption, flight, and heating behavior are responsible for 62% of the individually produced emissions. It also indicates that the average emission of the sample is lower than the Austrian average (9.4 tons). This may be caused by the sampling as the sample description shows that elderly people are overrepresented in

Table 5.1   Distribution of average GHG emissions by individual sectors

Sectors	CO_2 equivalents in kg	Percentage of total CO_2-eq
Housing		
Heating	750	10%
Warm water consumption	230	3%
Mobility		
Car use	1800	24%
Flights	1000	13.5%
Diet		
Meat consumption	1000	13.5%
Dairy products	640	9%
Other food[a]	580	8%
Consumption		
Car (purchase)	360	5%
Good consumption (clothing and electronic devices)	300	4%
Leisure activities	200	3%
Other[b]	500	7%
Total average	7360	100%

Source: OeNB sample (Hadler et al., 2021)

[a]"Other food" refers to the selection of food categories that did not belong into the "meat" and "dairy" category. These were, for example, coffee, fruits, nuts, sweets, vegetables, and so on.

[b]"Other" refers to other sectors included in the survey that contribute only little to the number of produced emissions and are therefore combined into one category, for example, the use of public transport, power consumption (household), or building information (household)

the survey. This will be further elaborated below, where the individual sectors and behaviors are described in detail.

Housing—Buildings, Heating, Water, Electrical, and Household Appliances

Most of our respondents live in single-family homes (45.7%), 44% in an apartment block or a high-rise building, and 10.3% in a semi-detached or terraced house. The average living space is around 128.69 m^2 (1385 sqft). A total of 42.6% of respondents state that their place is completely thermally insulated, and 37.8% state that it is only partially insulated. When asked about an energy performance certificate, only 28 respondents were able to show one; 71.4% of these certificates indicate at least class B or higher, which represents low energy or passive houses. The most frequently used main heating system is central heating (31.7%), 26.3% have a district heating system, and 18.5% a gas convector heating system. The most frequently stated main energy source is gas (36.9%), followed by district heating (23.6%), oil (18.2%), and wood/pellets/wood chips (15.8%).

The average heating costs are €112/month. On average, the heating within the sample results in CO_2 emissions of 1277.1 kg/year. Most emissions are caused by oil heating systems. The average emission here is 2489 kg CO_2/year, which is twice the sample's average. The average emission values of gas, district heating, and electricity are between 1232 and 1374 kg CO_2/year. The lowest CO_2 emissions are emitted by wood heating systems (116 kg CO_2/year) and heat pump/solar thermal systems (63 kg CO_2/year).

As for heating behavior, more than half of our respondents indicate that their heating is lowered when they leave their home for more than one day. If leaving for only one day, 31.1% lower their heating. Around 30% also say that they lower their heating at night. During the heating period, 47.8% of the respondents believe that their room temperature corresponds to the Austrian average. The actual temperature measured in the households by the interviewer using a thermometer was on average 22 °C (72 °F). The actually measured temperature was also compared with the subjective temperature assessment of the respondents. For 68.8% of the respondents, the estimated room temperature corresponded with the actually measured temperature—or was within the range of +/-1 °C.

Regarding water consumption, the respondents stated that they shower on average six times per week, and the average showering time is between five and seven minutes. Thirty percent also say that they take baths, with an average number of four baths per month. The most frequently used technology to heat water is a storage heater (52.2%), 27.8% say that they use an instantaneous heater or a gas-fired heater, and around 12.4% a solar system.

The total electricity consumption of the respondents is on average 4534 kWh/year. This value was calculated on the basis of those respondents who provided their electricity bill. The average electricity cost per month is around €131. The most commonly used household appliances are an electric cooker in the form of a ceramic hob, oven, microwave, dishwasher and washing machine. All these appliances are usually used at least several times a week. A total of 70.8% say they eat a hot meal six to nine times a week, and 46.9% cook most often for two people. This usage behavior of household appliances causes an average CO_2 emission of 105.2 kg CO_2/year.

The electrical appliances most commonly found in the households of the respondents are a television ($\leq$40 inch), a music or home cinema system, and a laptop. In full, 57.9% of the respondents say they do not own a large television set, 56.9% do not have a desktop computer in their household, and only 7.2% own an air conditioning system. Looking at the use of individual electronic devices, it can be seen that televisions are used on average between one and three hours per day. Desktops and laptops are used about two hours per day. Those with air conditioning indicate that they use it for one to two hours a day in the summer. In general, many of the respondents estimate their personal use of these electrical appliances as average (44%), and 30.6% think that their use is below average. This usage behavior causes an average CO_2 emission of 32.2 kg CO_2/year.

Mobility—Car Usage and Flight Behavior

On average, our respondents travel a distance of 5000–10,000 km per year in their most frequently used car. Data from the Verkehrsclub Österreich (VCÖ) show that the mobility behavior of our respondents is similar to that of the Austrian population. On average, Austrians travel around 9000 km per year by car (VCÖ, 2016), with commuting to work being a large share of these trips (see VCÖ, 2020a).

On average, our respondents estimate spending one and a half to two hours per week in the car. Around a quarter of the respondents stated that they spend more than 90% of the time alone in the car. Another quarter said they were "never or almost never" alone in the car. When they are not alone in the car, 70.2% say that there is one additional person besides the driver or passenger present. On average, car use causes CO_2 emissions of 1864 kg CO_2/year per person.

More than half of the respondents (57%) state that their most frequently used vehicle is a diesel car, 40.9% own a petrol-car and only two individuals (1%) an electric car. A total of 59.3% say that their car uses between 5 to 7 liters of fuel per 100 km and 26.6% estimate between 7 and 10 liters. On average, diesel vehicles emit 2319 kg CO_2/year, while petrol vehicles emit 1535 kg CO_2/year. This difference is mainly due to the fact that our respondents with diesel cars travel approximately 16,000 km per year, which is significantly above the average. For the two respondents who own electric cars, the average annual CO_2 emissions are 465 kg.

As for public transport, 37.7% of the respondents stated that they do not use public transport at all, 18.4% spend a maximum of 30 minutes on public transport per week, 18.9% between 30 minutes and two hours, and 14.9% more than five hours. Those who travel by public transport produce an average of 85.4 kg of CO_2/year as a result of this mobility. It should be noted that among those who use public transport, only 21.7% do not have a car.

When asked very generally about their flights, 43.3% of respondents said that they never fly abroad or only once every few years, 22.6% travel once a year, and 17.8% several times a year (both short- and long-haul flights). Asking about the preceding year specifically (which is referring to 2018–2019), 56.2% stated that they had not traveled by plane at all. In order to differentiate more precisely, a distinction was made between business and private flights and short- and long-haul trips. With regard to private flights, 64.1% did not take any short-haul flights in the previous 12 months, while 19.2% had taken two short-haul flights. The frequencies are even lower for private long-haul flights—89% report not having taken a private long-haul flight in the previous 12 months. When asked about business flights, it is also apparent that the majority did not take any short- (88.5%) or long-haul flights (95.7%). There are, however, a few respondents in the sample who took a large number of business flights.

Calculating the exact flight times, based on the requested information on the actual destinations of the mentioned flights, shows that most individuals (32.2%) spent between two-and-a-half and five hours on a plane, 30.6% spent more than 20 hours, and 22.1% spent between five and ten hours traveling by plane. The rest of the sample had spent either less than two-and-a-half hours on a plane or were in the midfield, with 10–20 hours the preceding year. The average CO_2 output of those who had flown in the previous 12 months is 1040.4 kg CO_2/year. A comparison of our sample with the Austrian average, however, shows that infrequent flyers are overrepresented in our sample. Data from VCÖ (2017) show that about half of the population takes flights once a year, whereas around one-third of the population does not take any flights. Furthermore, the most frequent trips are short-haul flights, which are also considered to be particularly polluting.

Diet—Meat and Dairy Products

Questions on nutrition included detailed assessments of the frequency of the consumption of energy-intensive foods (mainly animal products). Respondents were asked how often they consume, for example, sausage products, beef, pork, lamb, poultry, dairy products, fish, and seafood. Meat is most often consumed in the form of sausage products. Dairy products and eggs are also among the most commonly consumed animal foods. In total, 70.3% of the respondents eat sausage products up to three times a week, and 93.4% eat cheese and eggs up to three times per week. Among the types of meat, poultry is the most frequently consumed; 49.5% eat it one to three times a week. For beef and veal, it is 39.7% and for pork 37.8%. Fish and seafood are eaten by 38% one to three times a week. In sum, the meat consumption of our respondents causes on average 1005.6 kg CO_2/year. Only 4.4% of the respondents stated that they are vegetarian or vegan. Most respondents (57.8%) eat meat "in some meals," 18% "in most meals," and 19.4% "very rarely."

The reported diet of our respondents is somewhat lower than the official figures. Austria had an annual meat consumption of 62.6 kg per capita in 2019, which indicates that meat is consumed up to five times a week on average. In an EU comparison, Austria is in the middle in terms of consumption of various types of meat, but its consumption of pork is far above the European average (Statista, 2020b).

The questions on eating habits also included a query as to how often someone goes out to eat in restaurants and similar places. In all, 46.9% state that they eat out several times a month and 30.9% several times a year or even less often. Finally, when asked about the percentage of food thrown away in their household, 26.3% say that they do not waste any food, 45.4% say that they throw away a maximum of 5%, and 19% around 5–10%.

Consumption—Goods, Leisure Activities, and Information

The questions on consumer behavior are divided into the consumption of electronic goods, leisure activities, and clothing. The respondents' own assessment of their purchasing behavior in all three areas shows that they see themselves as rather "frugal" and "considerate." With regard to electronic goods, 74.8% of those surveyed say that they pay attention to "long use" and "only replace items when they are broken." Even when dealing with major household investments, 72.8% pay attention to the "longevity of products" when making purchases. When asked about clothing, the answers are more widely distributed, with 47.8% particularly emphasizing "long use," 31% rating their purchasing of clothing as "average," and 12.3% describing it as "very modest."

When asked about consumption of electronic goods, respondents were also asked to indicate, from a list of different items, how often they buy them new. Of all the electronics items listed, the smartphone is most frequently bought new. On average, a new smartphone is bought every three to five years, whereas a laptop is bought new after five to seven years and a television after six to ten years. A new car is bought on average less frequently than every ten years.

In the assessment of their personal leisure time behavior, 56.3% state that they need little equipment and infrastructure. They were also asked how often they buy new sporting equipment (e.g., bicycle and ski/snowboard). Almost half of the respondents (49.3%) buy a new bicycle less often than every ten years. The situation is similar when buying new skis or a new snowboard. Those who practice this winter sport buy new equipment on average less often than every ten years.

In addition to purchasing behavior, the leisure activities of the respondents were also examined in more detail. They were asked how often they had visited various leisure facilities in the previous 12 months. On average, the respondents most frequently visited a cinema/theater, an opera, or a football stadium (each three to five days a year). A ski resort was visited on average for one or two days, with more than half of the respondents not having visited one at all in the previous 12 months. More than 60% also say that they have never visited a theme park. Those who did were there for one or two days on average.

In addition, half of the sample (50.5%) had spent between 3 and 15 days in a hotel in the previous 12 months, 58% said they had not visited an apartment, bed and breakfast or youth hostel, and only 13% had spent six to ten days in one.

From a list of different items of clothing, the respondents indicated how often they had bought or received them as gifts in the previous 12 months. Of all the items given, a shirt was bought/gifted most often. On average, the majority of the respondents estimate that three to four shirts had been bought/gifted in the previous 12 months. The remaining items of clothing (shoes, trousers, skirts, sweaters, dresses, jackets, and coats) were most often bought/gifted one to two times a year. Converted to CO_2 consumption, this means, for example, an average of 93.6 kg CO_2/year for shirts and 106.7 kg CO_2/year for trousers and shoes.

5.3   Estimating and Explaining an Individual's Total Emission

The previous section discussed each emission area in detail. This section now uses a hierarchical linear regression model to identify sets of variables that allow for a good approximation of the respondents' total emission value. The goal is to explain as much variance as possible with the smallest number of variables. The identified small set of variables could then be used in future surveys that have only limited room for emission items.

Table 5.2 presents a model that is able to capture 77% of the emissions with five variables and an extended model that captures 88%. The first model includes variables that are associated with the largest quantities of GHGs in terms of content and can be surveyed validly, specifically the questions concerning the annual car-km, the consumption of beef and lamb products per week, the number of flights per year, the living space, and the number of people living in the household. The extended model includes additional variables, such as the amount of using a car alone, main heating source, long flights, consummation of pork and poultry per week, and number of shoes and phones per year as proxies for clothing and electronic devices. All the added variables in the second model have a significant influence on the explained emissions except for the main heating source.

Table 5.2   Estimating the total individual GHG emissions

	Total CO_2 equivalents			
	Model 1		Model 2	
Predictors	Estimates	std. Beta	Estimates	std. Beta
(intercept)	3591.93		1804.36	
Car kilometers	0.18	0.58***	0.16	0.51***
Beef and lamb per week	212.19	0.15***	189.46	0.13***
Number of flights per year	617.32	0.50***	295.19	0.24***
Living space in square meters	9.65	0.23***	9.39	0.22***
Number of household members	-636.63	-0.21***	-498.42	-0.16***
Percentage of using car alone			13.50	0.14***
Heating electricity			-73.44	-0.02
Heating gas			147.61	0.04
Heating oil			322.53	0.10
Heating solar			-599.39	-0.18
Heating wood			-936.98	-0.28***
Number of long flights			565.03	0.35***
Number of shoes per year			107.52	0.06**
Number of phones per year			2466.69	0.11***
Pork and poultry per week			183.53	0.10***
Observations	208		208	
R^2/R^2 adjusted	0.772/0.766		0.888/0.879	

*$p < 0.1$; **$p < 0.05$; ***$p < 0.01$. Linear regression. Source: OeNB sample (Hadler et al., 2021)

Factors that Shape the Total Emissions

Having shown how the emissions of an individual can be approximated with a few variables, we now want to provide some explanations for these emissions at the individual level. We start by adding the attitudinal and behavioral level in addition to relevant socio-demographic variables based on the theories and models discussed in Chap. 2. As noted earlier in the theoretical overview, explaining an individual's emission level requires not only the inclusion of individual social-structural factors but also the consideration of contextual factors such as geographical, political, or institutional settings. Including this contextual-level emission-oriented behavior has been well explained in past research (Kollmuss & Agyeman, 2002; Newton & Meyer, 2012; Stern, 2000; Tabi, 2013). This section, however, focuses on individual social-structural factors that have an impact on the emission-relevant behavior of individuals.

We will present the results of a regression that uses the total emissions as the dependent variable and a number of socio-demographic, attitudinal, and behavioral items as independent variables. We consider the following socio-demographics: income, age, education, gender, and place of residence[5] (see, among others, Huddart Kennedy et al., 2013; Gatersleben et al., 2002). In addition to these socio-demographic variables, attitudinal and behavioral intentions are also included since their significance is highlighted in past research (see, among others, Maloney & Ward, 1973; Gifford & Sussman, 2012). The items we use are taken from the ISSP (www.issp.org) and will be presented in detail in the following sections.

Dimensions of Environmental Attitudes and Behaviors

When asked "In general, how concerned are you about the environment?", 95.7% of the respondents say that they are "rather concerned" or "very concerned." The distributions of the individual items show that the majority of the sample has a positive attitude toward the environment and environmental protection. Based on the 22 questions on environmental attitudes and behavioral intentions, a total of six scales were considered, as follows: environmental concern, economic influence on the environment, influence of modern lifestyles on the environment, micro fatalism, acceptance of personal restrictions and climate policy measurements, as well as behavioral intentions toward environmentally oriented actions. Factorial analysis (VARIMAX) was used to determine the different dimensions.[6] The following sections summarize the composition of the scales and their distributions.

The scale "environmental concern" is intended to reflect the general concern about the environment and contains answers to the following questions: "How concerned are you about the environment?"; "There are more important things to do in life than protecting the environment"; "Many assertions about the threat to the environment are exaggerated"; and "We worry too much about the future of the environment these days and too little about prices and jobs." A low value on this scale means that the individual has little concern for the environment. The mean value is

---

[5] The place of residence can also be considered at the contextual level. It, for example, determines the mobility behavior due to infrastructural differences.

[6] Reliability analyses show a Cronbach's alpha between 0.547 and 0.639 within all six scales. All six scales have a range from 1 to 5.

4.2, and the mode is 4.75. This shows that a large part of the sample is highly concerned about the environment.

"Economic influences" is related to ecological modernization and reflects the attitude toward economic growth and its influence on the environment. It contains answers to the following questions: "Almost everything we do in our modern world harms the environment"; "In order to protect the environment, Austria needs economic growth"; and "Economic growth always harms the environment." A low value means that one is of the opinion that economic growth is not harmful to the environment. The mean value and median are around 3.4, and the skew value is close to zero. There is no clear tendency in the response behavior in any direction.

The scale "influence of modern lifestyles" indicates how the influence of modern everyday life on the environment is perceived. It contains answers to the following questions: "I find it difficult to judge whether my lifestyle benefits or harms the environment" and "Modern science will solve our environmental problems with little change in our lifestyle." A low value means that a respondent believes that a modern lifestyle has no negative impact on the environment. The mean value is 3.8, the median is 4, and there is a negative skew value. The majority of respondents therefore believe that modern lifestyles tend to have a negative impact on the environment.

The scale "micro fatalism" addresses subjective efficacy and contains answers to the following questions: "I do what is right for the environment even if it costs me more money or time" and "It is useless to do my part for the environment as long as others do not behave in the same way." A low value means you believe that your own behavior has no influence on the environment. The mean value is 4.2, and the skew value indicates a left-skewed distribution. Thus, a majority of the respondents reject these statements, which means that they believe in a positive influence at the microsocial level on the environment.

The scale "willingness to sacrifice for the environment" is closer to environmental behavior and contains answers to the following three questions: "To what extent would you find it acceptable … (a) … to pay much higher prices to protect the environment?; (b) … to pay much higher taxes to protect the environment?; and (c) … to sacrifice your standard of living to protect the environment?" This scale is intended to show the individual's willingness to accept personal restrictions and climate policy measurements to protect the environment. Therefore, it cannot be defined as actual behavior but more as an intention. A low value on the scale means that the proposed measures would be denied on a high level. The mean value (3.4) and median (3.3) are close, and the distribution indicates that a larger part of the sample shows approval of these measures.

Finally, we also consider environmental private-sphere behavior, which includes the following six questions: "How often do you do the following things? (a) separate valuable materials from your waste, such as glass, metal, plastic, paper, etc. for reuse (recycling); (b) buy fruit and vegetables that have not been treated with pesticides or chemicals; (c) limit driving for the sake of the environment; (d) reduce energy and fuel consumption at home for the sake of the environment; (e) save or reuse water for the sake of the environment; and (f) for the sake of the environment, avoid buying certain products." Since the respondent defines the answers to these questions as a mere self-declaration and thus the risk of socially desirable answers arises, they cannot be defined as actual environmental behavior. Rather, these answers are evaluated as behavioral intentions and are part of Stern's (2000) private-sphere behavior that has a small impact on the environment. The scale thus represents the intention of an individual to perform the above environmentally oriented actions. A low value on the scale means that an individual has a low behavioral intention and does not indicate taking these actions often. The mean value and median are around 3; a large part of the sample thus shows a higher behavioral intention to take the above actions.

Factors that Shape the Total Emissions

Table 5.3 shows the results of a linear regression with the total emissions as the dependent variable and the above-mentioned socio-demographics and scales as independent variables. The model explains a quarter (25.9%) of the dispersion of the total produced emissions of an individual. It shows that the attitude variables have no significant influence on the individual emission consumption. This is somewhat expected since past studies have pointed out that attitudes are important regarding the intention to change behavior but have less influence on actual behaviors (Abrahamse & Steg, 2009). Only the willingness to make sacrifices for the environment was significant. The negative beta-coefficient of the willingness scale (-0.15) indicates that individuals who have a high willingness to accept these restrictions for the environment are also more likely to produce less emissions. This suggests that there are some individuals who already produce

Table 5.3 Linear regression model; dependent variable: total CO_2 equivalents

Predictors	Total CO_2 equivalents	
	Estimates	Std. Beta
Intercept	7299.26***	
Residential area [ref. urban][b]		
Countryside	1221.89	0.43*
Suburban	1489.28	0.53**
Gender [ref. male]		
Female	-708.17	-0.25
Education [ref. university]		
(no) compulsory school	-923.18	-0.33
Apprenticeship	57.02	0.02
Vocational school (BMS)	-654.44	-0.23
Upper secondary school (AHS/BHS)	-104.80	-0.04
Age	-54.58	-0.35***
Income	1.55	0.52***
Income not reported[a]	1844.67	0.28**
Attitudes and intentions		
Environmental concern	-169.96	-0.04
Economic influence	-98.77	0.03
Influence of modern lifestyles	82.32	-0.03
Micro fatalism	228.52	0.05
Willingness to sacrifice for the environment	-495.73	-0.15*
Environmental private-sphere behavior	-738.18	-0.13
Observations	194	
R^2/R^2 adjusted	0.321/0.259	

$*p < 0.05$; $**p < 0.01$; $***p < 0.001$

[a]A dummy variable was included for respondents who did not report their income and interacted with the main term. In our case, respondents who did not report their income consume 1844.67 CO_2 equivalents more than reported in the intercept. "Income" is the linear effect per unit income, which is per euro. Source: OeNB sample (Hadler et al., 2021)

[b]Additionally, further differentiations within the residential areas were made, that is, differences within the two largest cities in the sample (Vienna vs. Graz) and differences between federal states (Vienna and Lower Austria vs. Styria). There were no significant effects.

less emissions based on their behavior and are also willing to make more sacrifices for the environment.

The socio-demographic variables of age, income, and residential area have significant effects. Younger respondents, urban dwellers, and respondents with a higher income produce more emissions. Looking at the standardized beta-coefficients in the model, the socio-demographics have the strongest influence on emissions (values over 0.3). These findings are thus in line with previous studies, which found a strong influence of these factors on an individual's consumption (Poortinga et al., 2004).

5.4   CONCLUSIONS AND OUTLOOK

This chapter started with a comparison and description of the emissions of Austrians and the study's respondents, which showed that our sample has somewhat lower CO_2 emissions than the average Austrian. The comparison of different areas of emissions made clear that most emissions are caused by car use, meat consumption, and flight behavior. The use of a regression model showed that five key variables are sufficient to estimate around three-quarters of the total CO_2 emissions caused by an individual. This finding suggests that asking about annual car kilometers, consumption of lamb and beef per week, number of flights, size of the living space, and number of household members allows for a quick assessment of an individual's CO_2 footprint.

As for factors shaping the CO_2 output, around a quarter of the dispersion can be explained by socio-demographics and willingness to act environmentally. Especially socio-demographic variables such as age, income, and residential area are strong and significant influences. Also, the willingness of someone to accept restrictions has a significant influence on emissions. The considered attitude scales had no significant effect.

What savings and guidelines for action can be derived from these results? Considering the areas that produce the most emissions, mobility, diet, and housing need to be addressed. As for buildings, Austria offers a number of subsidies, which vary from one federal state to other.[7] However, subsidy guarantees for private homes and other forms of private housing have been declining every year since 2014. In 2018, subsidy expenditure was almost -18% below the average of the previous ten years. There were also returns of 40% in 2018 compared to 2010 for renovation subsidies, putting Austria in the bottom third of the European subsidy expenditure.

---

[7] For example, subsidies for private homes, multi-story residential buildings, ecological requirements ("Ökoförderungen"), residential building checks ("Wohnbauscheck"), or residential building renovations. These subsidies are tied to a wide range of requirements, which, among other things, require the implementation of ecological measures by the developer (Land Steiermark, 2020).

At this level, a tripling of the renovation rate is necessary to meet the climate targets (IIBW/FV Steine-Keramik, 2019).

Another major contribution to national emissions is made by individual passenger transport. The VCÖ Mobility Survey (2020) shows that more than half of Austrians (57%) would prefer to cover some of the distances they have traveled by car by alternative means of transport (e.g., public transport, bicycle, and walking). At the same time, it is also apparent that, particularly in peripheral districts, the public infrastructure (e.g., cycle paths and railway stations) is considered to be insufficient (VCÖ, 2020b). The demand for alternative possibilities to the car is therefore given. Hence, transport policy measures should place a focus on the expansion of the cycle network and public transport in areas outside of large cities in order to offer those residents more alternatives to car use. Promotion of alternative mobility concepts such as car sharing[8] or bike sharing[9] can also contribute to reducing emissions, especially in peripheral areas.

As expected, the number of flights per year contributes to a high carbon footprint. According to the VCÖ (2020c), particularly short-haul flights are extremely harmful to the environment. In 2019 around five million passengers traveled by short-haul flights in Austria, each time covering distances less than 800 kilometers, which is equivalent to a flight duration of two hours or less. Therefore, the focus should be on reducing the number of these flights by either enforcing a higher tax or finding alternatives to reduce the necessity of these short flights, such as online meetings as alternatives to business trips or the use of alternative transportation such as railways or buses.

One last emission-intensive area would be the consumption of food, especially the consumption of animal products. It is a long and emission-intensive way until the meat lands on someone's plate. Looking at it from a consumer-based approach, it is important to raise awareness as to what sacrifices are made on different ecological levels when it comes to meat production and consumption. Generally, food, as a huge part of an individual's consumption, should be guided by environmentally conscious

---

[8] This mobility concept refers to the rental of a car. A distinction is made between providers with fixed locations and free-floating systems (no fixed location, borrow and return within a defined zone) (Stadt Wien, 2020). An example would be the mobility concept called "tim" in Graz, which does include the concept of car sharing among other alternative mobilities.

[9] This mobility concept refers to the rental of bicycles within a public space. There is also a distinction between stationary-based and free-floating systems. An example in Austria would be "city bike" in Vienna.

decisions to make up for the number of emissions that are caused through consumption.

The CO_2 quantities calculated in the study illustrate which areas of the household and personal behavior can be defined as emission-intensive. In addition to the already known "emission sinners," such as meat consumption and car use, the calculations also showed where savings can be made on a smaller scale. An essential area in households is water treatment and water consumption. The average CO_2 emissions here are 228.3 kg CO_2/year. In comparison to electricity consumption, the CO_2 emissions caused by water consumption are higher. Thus, water-saving measures can make a small but important contribution to reducing CO_2 consumption within a household. The usage behavior of the electrical appliances surveyed (e.g., TV, computer, and laptop) causes an average CO_2 emission value of only 32.2 kg CO_2/year, which is only 0.4% of the total emissions caused. This is noteworthy since some of these behaviors are also interpreted as energy-saving behavior—for example, energy-saving recommendations regarding the standby consumption of electrical appliances. These figures indicate that the use of everyday electrical appliances produces fewer emissions than suggested by the energy-saving recommendations and that the focus needs to be shifted.

In sum, this chapter pointed out which areas of life are particularly CO_2 intensive and which individual factors influence total emission output. The following chapter will add to this perspective by identifying specific patterns of consumption. It will show that there are specific lifestyles that are characterized by high energy demands in only one of the six sectors of consumption.

References

Abrahamse, W., & Steg, L. (2009). How do socio-demographic and psychological factors relate to households' direct and indirect energy use and savings? *Journal of Economic Psychology, 30*, 711–720.

Gatersleben, B., Steg, L., & Vlek, C. (2002). Measurement and determinants of environmentally significant consumer behavior. *Environment and Behavior, 34*(3), 335–362.

Gifford, R., & Sussman, R. (2012). Environmental attitudes. In S. D. Clayton (Ed.), *Oxford library of psychology. The Oxford handbook of environmental and conservation psychology* (pp. 65–80). Oxford University Press.

Hadler, M., Schweighart, M., & Wardana, R. (2021). *OeNB CO2-relevant environmental behavior*. Data will be available for free at the Austrian Social Science Data Archive. www.aussda.at; https://doi.org/10.11587/WQGMKY

Huddart Kennedy, E., Krahn, H., & Krogman, N. T. (2013). Are we counting what counts? A closer look at environmental concern, pro-environmental behavior, and carbon footprint. *Local Environment: The International Journal of Justice and Sustainability, 20*(2), 220–236.

IIBW/Steine-Keramik. (2019). *Wohnbauförderung in Österreich 2018*. Novographik Druck GmbH.

Kollmuss, A., & Agyeman, J. (2002). Mind the gap: Why do people act environmentally and what are the barriers to pro environmental behavior? *Environmental Education Research, 8*(3), 239–260.

Land Steiermark—Amt der Steiermärkischen Landesregierung. (2020). *Förderungen*. https://www.wohnbau.steiermark.at/cms/beitrag/12763938/155599590/

Maloney, M. P., & Ward, M. P. (1973). Ecology: Let's hear from the people: An objective scale for the measurement of ecological attitudes and knowledge. *American Psychologist, 28*(7), 583–586.

Newton, P., & Meyer, D. (2012). The determinants of urban resource consumption. *Environment and Behavior, 44*(1), 107–135.

Poortinga, W., Steg, L., & Vlek, C. (2004). Values, environmental concern, and environmental behavior. *Environment and Behavior, 36*(1), 70–93.

Stadt Wien. (2020). *Was ist carsharing?* https://www.wien.gv.at/verkehr/kfz/carsharing/wissenswertes.html

Statista. (2020a). *Treibhausgasemissionen in Österreich*. https://de.statista.com/statistik/studie/id/59925/dokument/treibhausgasemissionen-in-oesterreich/

Statista. (2020b). *Konsum von Fleisch in Österreich*. https://de.statista.com/statistik/studie/id/32545/dokument/konsum-von-fleisch-in-oesterreich-statista-dossier/

Stern, P. C. (2000). New environmental theories: Toward a coherent theory of environmentally significant behavior. *Journal of Social Issues, 56*(3), 407–424.

Tabi, A. (2013). Does pro-environmental behavior affect carbon emissions? *Energy Policy, 63*, 972–981.

Umweltbundesamt. (2020). *Klimaschutzbericht*. Wien. https://www.umweltbundesamt.at/studien-reports/publikationsdetail?pub_id=2340&cHash=04535f1c207c6ac8814ee0edb3809750

VCÖ. (2016). *VCÖ: Auto hat für Mobilität der Österreicher geringere Bedeutung als im EU-Schnitt*. https://www.vcoe.at/presse/presseaussendungen/detail/vcoe-auto-hat-fuer-mobilitaet-der-oesterreicher-geringere-bedeutung-als-im-eu-schnitt

VCÖ. (2017). *VCÖ: Jeder dritte Österreicher fliegt nie—jeder sechste fliegt mehrmals im Jahr.* https://www.vcoe.at/news/details/vcoe-jeder-dritte-oesterreicher-fliegt-nie-jeder-sechste-fliegt-mehrmals-im-jahr

VCÖ. (2020a). *Infographiken Mobilität allgemein.* https://www.vcoe.at/publikationen/infografiken/infografiken-mobilitaet-allgemein

VCÖ. (2020b). *VCÖ-Factsheet 2020-04—Österreichs Bevölkerung ist sehr vielfältig mobil.* https://www.vcoe.at/publikationen/vcoe-factsheets/detail/vcoe-factsheet-2020-04-oesterreichs-bevoelkerung-ist-sehr-vielfaeltig-mobil

VCÖ. (2020c). *VCÖ: Kurzstreckenflüge verstärkt durch Bahn und Videokonferenzen ersetzen.* https://www.vcoe.at/presse/presseaussendungen/detail/vcoe-inlandflug-in-oesterreich-verursacht-50-mal-so-hohe-klimaschaedliche-treibhausgase-wie-bahnfahrt

Windsperger, A., Windsperger, B., Bird, D. N., Jungmeier, G., Schwaiger, H., Canella, L., Frischknecht, R., Nathani, C., Guhsl, R., & Buchegger, A. (2017). *Life cycle based modelling of greenhouse gas emissions of Austrian consumption. Final report of the research project ClimAconsum to the Austrian Climate and Energy Fund, Vienna.* Institut für Industrielle Ökologie (IIÖ).

Windsperger, B., Windsperger, A., Bird, D. N., Schwaiger, H., Jungmeier, G., Nathani, C., & Frischknecht, R. (2019). Greenhouse gas emissions of the production chain behind consumption of products in Austria. Development and application of a product- and technology-specific approach. *Journal of Industrial Ecology, 1–12.*

CHAPTER 6

The Multidimensionality of Consumption: Energy Lifestyles

The previous chapter concluded by considering the factors that shape the total greenhouse gas (GHG) emissions of our respondents. The current chapter[1] extends this view by explicitly considering the multidimensionality of behavior. It considers the energy demand in the considered six areas of life (housing, mobility, consumption of goods, diet, leisure activities, and information) and combines them into unique patterns of energy consumption—that is, the "Energy Lifestyles" of the Austrian population.

It thus enhances the previous understanding of Energy Lifestyles in Austria in a specific way. While some earlier studies of survey data obtained implausible or inconsistent results by drawing conclusions from the respondents' lifestyles (based on psychological characteristics) on their (energy) behavior (Hierzinger et al., 2011; Bohunovsky et al., 2011), the current analysis starts from the energy demands of the respondents—that is, energy-related lifestyles are identified by clustering respondents according to their annual primary energy demands in the different areas of social life.

By identifying and analyzing a number of distinctive Energy Lifestyles, this chapter provides an overview of the quality and quantity of energy-related behavioral patterns in Austria. It shows that the overall lifestyle-related energy demands of the identified groups are composed in entirely

---

[1] Lead author: Stephan Schwarzinger. This chapter is based on the lead author's doctoral thesis.

© The Author(s) 2022
M. Hadler et al., *Surveying Climate-Relevant Behavior*,
https://doi.org/10.1007/978-3-030-85796-7_6

different ways and that an "average Energy Lifestyle" is the exemption. Using this classification as a basis, it also discusses target group-oriented policy interventions.

6.1   IDENTIFYING LIFESTYLES BASED ON ANNUAL ENERGY DEMANDS

The Energy Lifestyle framework used here (Schwarzinger et al., 2019b) is based on the three-part lifestyle concept by Lüdtke (Lüdtke, 1996). The first part, *Performance*, focuses on the observable facts about an individual's behavior, practices and relations to their physical surroundings. In the case of Energy Lifestyles, specific patterns of "Performance" are the reason for individuals to have specific patterns of energy demand across different areas of life. Groups of individuals with similar Energy Lifestyles can then be identified by clustering people according to their estimated energy demand patterns. The second part, *Situation*, represents the objective context in which a lifestyle-specific behavior is conducted. It contains information on socioeconomic characteristics, cultural resources, and constraints. The third part, *Mentality*, focuses on psychological characteristics such as perceptions, values, and preferences. Methodologically, "Performance" is the basis for identifying and describing lifestyle groups' characteristic behaviors (and related impacts), while "Situation" and "Mentality" are essential for achieving a broader understanding about how different lifestyle groups live. This behavior-centered perspective aligns relatively well with understanding a lifestyle as a "system of classified and classifying practices" (Bourdieu, 1987).

Why is clustering respondents on the basis of estimated energy demands considered a more appropriate approach to identifying Energy Lifestyles than clustering them according to their psychological or socio-demographic characteristics? A comparison of different approaches to group identification, using the same representative dataset from 2009, showed that the former method leads to more useful results than clustering respondents according to psychological or socio-demographic characteristics (Bohunovsky et al., 2011; Schwarzinger et al., 2018). This becomes particularly relevant when groups with distinct energy behavioral patterns shall be identified and in a second step characterized by psychological and socio-demographic variables.

As indicated above, the Energy Lifestyle concept used here (Schwarzinger et al., 2019b) distinguishes six areas of life in which individuals can behave according to their preferences and make choices with regard to the technical equipment they use or purchase ("Performance"). The six areas are "housing," "mobility," "diet," "consumption," "leisure," and "information." In each of these areas, a variety of climate- and energy-relevant goods and services can be consumed. They represent "dimensions of energy consumption."[2] To identify groups of people with similar Energy Lifestyles, individuals are clustered according to their patterns of energy consumption across these dimensions. But why are energy demands used as a proxy for energy- and climate-relevant behavior instead of GHG emissions? A person's behavior results in a certain energy demand. The link between behavior and energy demand is practically deterministic. For example, a respondent's car consumes a certain amount of fuel per distance, and the respondent drives a certain number of kilometers per year, accompanied by a certain number of people. The total amount of fuel and (under consideration of the whole supply chain of the respective fuel type) the amount of primary energy used can be calculated on the basis of these numbers by life-cycle assessment (LCA) methodology (Finkbeiner et al., 2006). When all the collected responses about behavior and equipment use are processed in this way, an estimate of each respondent's annual primary energy demand in each of the six areas of life can be obtained. These *estimated annual primary energy demands* (in the following referred to as simply "energy demands") appear as reasonable indicators for the energy intensity of a person's behavior in each of the six dimensions of energy consumption (or areas of life). Finally, lowering the climate impact of these lifestyles can be achieved by either reducing the behavior or generating carbon-neutral energy.

This operationalization of Energy Lifestyles also implies certain limitations. The approach was explicitly developed for the "bottom-up" group-level assessment of Energy Lifestyles and not as a substitute for top-down national statistics. For example, building or buying a home is often an important part of an individual lifestyle. However, (unlike, e.g., for cars) there is currently no reasonable way to estimate the energy demand for the construction of dwellings on an individual level as there is little

---

[2] Due to limited personal choice, the energy demands and emissions related to public infrastructure, emergency services, or the reception of medical treatment are not considered part of a person's lifestyle.

information available on the lifespan of buildings. Thus, the methodology focuses on lifestyle-specific behavior where the associated energy demand can be attributed to with reasonable accuracy (see also Chap. 3).

6.2   Data and Methods

The dataset used in this study is based on a representative sample of the Austrian population consisting of 604 respondents. It was collected in the course of the Horizon 2020-funded project "ECHOES" in summer 2018 (together with samples from 30 other European countries) in an online-survey setting in which respondents received a small monetary reward for their participation (Reichl et al., 2019). Among other tasks, a cross-European analysis of Energy Lifestyles was carried out with these data applying the same method of group identification as that used herein (Schwarzinger et al., 2019a).

The survey covered the three main lifestyle components of "Performance," "Situation," and "Mentality" (Lüdtke, 1996). With regard to "Performance" and "Mentality," the survey had a particular focus on energy behavior in order to enable both an LCA estimation of the energy demand resulting from individual behavior and a discussion of the role of psychological characteristics with regard to behavioral patterns. The raw dataset and questionnaire are available online[3] (Reichl et al., 2019).

- *Behavior* (*"Performance"*) was covered by questions related to energy- and climate-relevant behavior and equipment use in the six areas of life. The items were designed in such a way that they could be answered by normal users and consumers without special technical knowledge.
- With regard to *socio-demographics* (*"Situation"*), age, sex, urban/rural residence, household size, number of children under 14, educational level, and subjective social status were used.
- *Psychological characteristics* (*"Mentality"*) were covered by variables and scales on political orientation, beliefs regarding the implications of the spread of renewable energy sources, belief in climate change, perceived normative pressure, self-efficacy, subjective personal obligation, pro-environmental identity, intention to support the energy transition, and acceptance of energy policy measures.

---

[3] https://db.echoes-project.eu/echoes/home

Additionally, the Energy Lifestyle framework (Schwarzinger et al., 2019b) considers "Context," which stands for specific circumstances that are associated with the country or region a respondent lives in. In the case of Austria, ZIP codes were used to define seven regions. While socio-demographic and psychological variables could be used without any uncommon transformations, data on behavior and equipment use had to be processed into estimations for the respondents' energy demands in the six areas of life.

As described in Chap. 3, an LCA-based estimation of energy demands on the basis of self-reported behavior and equipment use ("Performance") was used. It had already been conducted in the course of a cross-European study of Energy Lifestyles during the ECHOES project (Schwarzinger et al., 2019a). It was estimated separately for all surveyed activities and then summarized in the six areas of "housing," "mobility," "diet," "consumption," "leisure," and "information" (Bird et al., 2019). As the LCA estimations were made under consideration of country-specific conditions (e.g., heating/cooling degree days), they could be used for the Austria-specific analyses without the need for further modifications.

6.3   Multidimensional Identification of Energy Lifestyles

According to the Energy Lifestyle framework (Schwarzinger et al., 2019b), the annual individual energy demand in a certain area of life is a result of behavior conducted in the respective field. The estimated energy demands represent the energy intensity of activities carried out by an individual in the course of a year (as far as they are covered by the survey). In the following, the results of clustering respondents on the basis of estimated energy demands in six areas of life are presented. Socio-demographic-, psychological-, and context-related variables are used to characterize the identified groups and to obtain a holistic impression of the groups' Energy Lifestyles.

Before considering the results in detail, a short look at the aggregate level is helpful. The data represent the energy impacts associated with the lifestyle of an "average Austrian." On average, with around 39,000 and 38,000 megajoules (MJ) per year, respectively, housing and mobility are head-to-head and together account for around 70% of an average Austrian's annual energy impact. Diet accounts for around 15,000 MJ, leisure activities for 13,000, consumption of durable goods and clothing for 6000, and the acquisition of information for less than 1000 MJ per year. At this point in time, the distribution of primary energy impacts is closely linked to the distribution of emissions. This is due to the fact that technologies using fossil fuels achieve low overall energy efficiency and at the same time are emission intensive. However, there is an important reason for the current focus on Energy Lifestyles instead of "Emission Lifestyles"—"Emission Lifestyles" cannot be identified once true decarbonization is achieved (which shall be the case by around 2050). Instead, Energy Lifestyles can continue to be analyzed even when a "zero carbon economy" has been established. Lifestyles based on emissions, however, will be presented in Chap. 8.

The respondents were clustered on the basis of each respondent's estimated energy demand in each of the six areas of life (housing, mobility, consumption, diet, leisure, and information). As indicated above, the energy demands in the different areas show very different scales. Therefore, before being passed to a conventional and well-documented k-means clustering algorithm, they were z-transformed. This procedure minimized the arbitrary and systematic influence on the outcome. A solution with five Energy Lifestyle clusters was chosen. It provides a reasonable balance between interpretability and level of detail for a nation-wide overview. Figure 6.1 depicts the energy demand patterns of the five groups in relation to the Austrian average (100% baseline).

The energy demands of the five groups are composed in very different ways. Lifestyle 1 is characterized by a peak in the mobility dimension. Lifestyle 2 is below average in all dimensions. Lifestyle 3 is above average in the two dimensions leisure and information, lifestyle 4 in housing, and lifestyle 5 in consumption. For example, lifestyles 3 and 5 have overall energy demands virtually equal to the national average of about 113,000 MJ per year, but their patterns differ from each other and from "the average Austrian." Apparently, group profiles represent different variations of the expressive component ("Performance") of energy-related lifestyles.

What also becomes particularly clear in this figure is that the areas of life with the greatest mean energy demand (housing and mobility) are also the areas with the greatest relative differences between the groups. Diet, in contrast, shows a remarkably small variation between groups. This is an indication that the type (and thus the energy intensity) of an individual's diet barely corresponds with energy-related behavior in other areas of life.

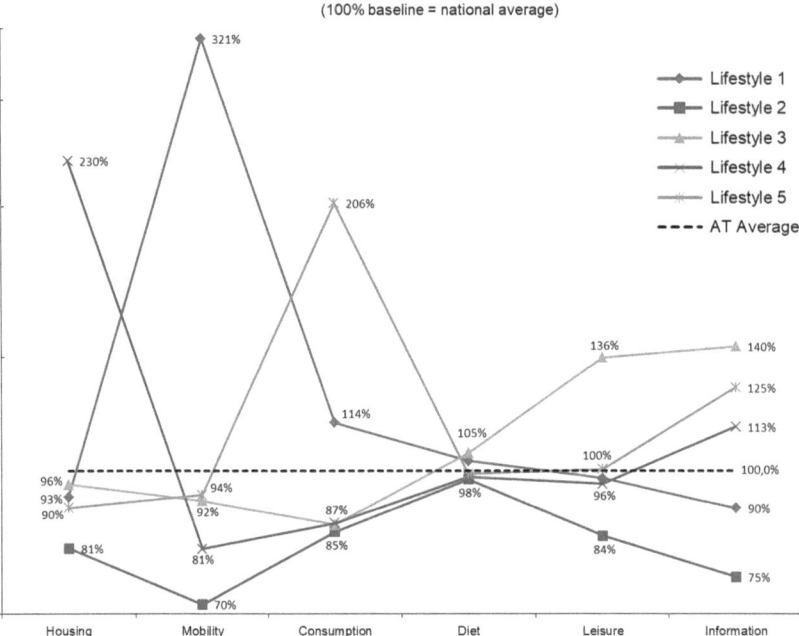

Fig. 6.1 Energy demand profiles of five Energy Lifestyle groups in Austria. (Source: Calculation based on ECHOES data [Reichl et al., 2019])

Consumption, the area of life with the fourth largest average share of individual overall primary energy demand, shows relatively large differences between the groups, while leisure as well as information show slightly smaller differences between groups.

6.4   ENERGY LIFESTYLE GROUPS AND THEIR ROLE IN THE ENERGY TRANSITION

Table 6.1 provides a summary of the overall energy demand of each lifestyle group as well as information on the manifest living conditions ("Situation" and "Context") and the perceptions, norms, and beliefs of respondents in the identified groups ("Mentality"). These characterizations are also based on correspondence analyses and provide various

Table 6.1 Energy Lifestyle groups and their specific characteristics

	Lifestyle 1 "Travelers"	Lifestyle 2 "Savers"	Lifestyle 3 "Hobbyists"	Lifestyle 4 "Homers"	Lifestyle 5 "Consumers"
Size of group	8.6%	48.8%	23.5%	8.9%	10.1%
Overall energy use	195,562 MJ	89,954 MJ	112,944 MJ	155,076 MJ	112,273 MJ
Regional context	Upper Austria overrepresented	Vienna/Lower Austria, Carinthia and Salzburg overrepresented	Vienna/Lower Austria, Carinthia and Salzburg overrepresented Urban areas overrepresented	Vorarlberg/Tyrol and Styria overrepresented Rural areas overrepresented	Burgenland and Styria overrepresented Urban areas overrepresented
Sex	Men overrepresented	Women overrepresented	Men overrepresented	Women overrepresented	Women overrepresented
Age groups	18–34 underrepresented, 35–54 overrepresented	18–34 underrepresented, 45–54 and 55+ overrepresented	18–34 overrepresented, 45–54 and 55+ underrepresented	18–34 and 35–44 underrepresented, 55+ overrepresented	18–34 overrepresented, 35–54 underrepresented
Household size	Often 2–5	Often 2–5	Often >5	Often 1	Often 3, 4, or 5+
Education	Higher categories	Diverse	Diverse, lower categories	Diverse	Diverse, higher categories
Subjective social status	Higher categories	Lower categories	Average	Diverse, lower categories	Higher categories
Labor market integration	Good	Limited labor market integration	Good	Limited labor market integration	Good, elevated share of full-time students

Source: Calculation based on ECHOES data (Reichl et al., 2019)

insights into the lifestyles typical for the respective groups. Moreover, they provide potential starting points for group-specific policy measures toward a lower energy- and climate-related impact.

Energy Lifestyle 1: "Travelers"

Energy Lifestyle 1 is characterized by an average overall energy demand of approximately 196,000 MJ per year. A total of 8.6% of all Austrian survey respondents were assigned to this group on the basis of their patterns of energy demand. Respondents from Upper Austria are overrepresented in this group. Men and people between the ages 35–54 are overrepresented, while respondents between 18 and 34 are underrepresented. There is a tendency toward larger households and large living spaces per capita. Together with an above-average share of fossil-fueled heating systems, this results in the third highest energy demand for "housing." A tendency toward higher education can be observed. Right political orientation is overrepresented. Furthermore, there is a tendency toward a higher subjective social status. Many respondents report good labor market integration by being either self-employed or employed 30 hours per week or more. In "mobility," this group is characterized by an outstandingly high energy demand of 320% of the Austrian average. This is a result of extraordinarily long annual driving distances, much air travel and little use of public transport. Although the share of hybrid-electric and fully electric cars is higher than in the Austrian average, their share is not high enough to result in a considerable decrease in mobility-related energy demand on the group level. The group is partially interested in car sharing or has already tried and liked it. However, the question arises as to how well car sharing is compatible with the tendency toward high annual driving distances. Regular bicycle use is not very common, whereas the proportion of respondents who use the bicycle "sometimes" is elevated. The slightly above-average energy demand for "consumption" is a result of a tendency toward higher consumption activity in both fashion and electronics. Despite some meatorientation in the group-specific dietary preferences, these preferences are not so common as to lead to an elevated energy demand. Regarding leisure activities, there is an above-average share of activities that require a moderate amount of equipment and infrastructure, while activities with more extensive needs are underrepresented. Although there is a tendency toward an increased use of electronics, the group has a below-average energy demand for "information." This corresponds well with a high share of electricity from renewable energy production.

Overall, this group generally shows a tendency toward intensive consumption and energy-intensive practices. Technically, there are potentials for modernization (e.g., the current overrepresentation of fossil-fueled heating systems) and indications for interest in sustainable technologies (e.g., overrepresentation of hybrid-electric and electric vehicles). On the one hand, disbelief in climate change and small scores in environmental identity as well as in the perceived normative pressure to act in accordance with the energy transition are more frequent than in the Austrian average. Additionally, the intention to support the energy transition is small. On the other hand, lifestyle 1 respondents have a tendency toward high self-efficacy with regard to the energy transition and often strongly agree with the statement that renewables will create new jobs. With regard to enabling this group toward the energy transition, intrinsic factors do not appear to be a suitable starting point for policy interventions. By contrast, linking the energy transition with the consumption of innovative technologies and related economic opportunities might be a more suitable strategy.

Energy Lifestyle 2: "Savers"

Energy Lifestyle 2 is characterized by an average overall energy demand of approximately 90,000 MJ per year. In total, 48.8% of all Austrian survey respondents were assigned to this group on the basis of their patterns of energy demand. Respondents from Vienna/Lower Austria, Carinthia, and Salzburg are overrepresented in this group, while Styria and Burgenland are underrepresented. Women and age groups 45 and above are overrepresented, whereas respondents between 18 and 34 are underrepresented. Similar to lifestyle 1, there is some tendency toward larger households but with smaller living space categories. This, in combination with a higher share of modern heating systems, results in the lowest group-specific energy demand for "housing." Regarding education, there is a diverse overall picture in the group. Compared to the Austrian average, respondents show a tendency toward reduced labor market integration and often describe their social status as "below average."

In the area "mobility," the group also has the lowest energy demand of all groups (69.7% of the Austrian average). This reflects their small annual driving distance and an above-average use of public transport. Furthermore, it reflects a lower number of private flights. The picture with regard to car sharing is diverse, with some indication of interest in this subject. In this context, it might be interesting to determine which individuals are particularly open and interested. If car sharing allows people to give up owning a car, this can be accompanied by energy savings and ecological benefits. However, if trips that are currently made by public transport are shifted to the car as a result of car sharing, the ecological effect is likely to be negative. As in lifestyle 1, regular bicycle use not very common. Also in the area of "consumption," lifestyle 2 group members have the lowest primary energy demand of all groups. This reflects a tendency toward modest consumption behavior in both fashion and electronics.

Although there is an above-average share of respondents with reduced meat consumption, the dietary preferences and impacts as a whole are again hardly group-specific. The leisure activities of respondents assigned to this group cause the lowest energy demand of all groups. Activities that require limited amounts of equipment and infrastructure are overrepresented. With regard to electronics, there is a tendency toward low usage in combination with an overrepresentation of "green electricity," which is reflected in the fact that the group also has the lowest average energy demand for "information."

This group accounts for nearly half of the sample. It has the lowest overall energy demand of all identified groups and a below-average energy demand in all six areas of life. A relatively large share of respondents has a high self-efficacy with regard to the energy transition, a pro-environmental self-identity, and the intention to support the energy transition. Many in the group see it as a personal obligation to be energy efficient and to behave in accordance with the energy transition. Many would accept policy measures that result in higher costs. However, social status and labor market integration indicate that the small energy impact among this group is not only a consequence of conscious decisions regarding energy use but also a result of economic restrictions. In addition to the aforementioned tendency toward pro-environmental positions, there is also an above-average rejection of the statement that renewables are good for the environment, which might be an indication of the existence of subgroups with diverging positions.

Although the group has very low energy demands, the nation-wide potential of a more systematic implementation of energy-efficient technologies within lifestyle 2 should not be underestimated due to the large size of the group. In order to develop target-oriented policy measures (e.g., funding for the acceleration of costly technological upgrades), a more detailed exploration within the group seems helpful because the information at hand leaves relatively much room for speculation.

Energy Lifestyle 3: "Hobbyists"

Individuals with Energy Lifestyle 3 have an average overall energy demand of approximately 113,000 MJ per year. In total, 23.5% of all Austrian survey respondents were assigned to this group on the basis of their patterns of energy demand. Respondents from Vienna/Lower Austria, Carinthia, and Salzburg are overrepresented in the group, while Styria, Burgenland, and Upper Austria are underrepresented. Men and ages 18–34 are overrepresented, whereas respondents of 45 years and above are underrepresented. There is a tendency toward large households of more than five people and smaller living space categories. This, in combination with a higher share of fossil-fueled heating systems and an overrepresentation of blocks with more than ten dwellings, results in a "housing"-related energy demand that is slightly below the national average. Regarding education, the overall picture is diverse, with some tendency toward lower levels. Left political orientations are overrepresented. Compared to the Austrian average, respondents tend to show good labor market integration and often see their social status as "average."

Regarding "mobility," this group is characterized by a below-average energy demand. This reflects the high share of close-to-average annual driving distances and a diffuse overall picture with regard to air travel. There is a tendency toward low usage of public transport and little interest in car sharing. The group shows a diverse picture regarding bicycle use but also an above-average share of all-season cyclists. The below-average energy demand for "consumption" results from a tendency toward reduced consumption activity in fashion and an above-average share of respondents who "like to always have the latest technology" in electronics. Despite an overrepresentation of meat-oriented respondents, the "diet"-related energy demand is again close to the average. Regarding leisure activities, there is an above-average share of activities that require an elevated to high amount of equipment and infrastructure, which leads to the highest average energy demand for "leisure" of all five groups. There is a tendency toward an intensive use of electronics. The above-average energy demand for "information" reflects the combination of an intensive use of electronics and a below-average share of respondents who reported using "green electricity."

In sum, this group is practically average with regard to the overall primary energy demand. The distribution across the six areas of life shows, however, an elevated energy demand for leisure activities and use of electronic means for information acquisition. Although this group appears relatively unremarkable with regard to energy impacts at first glance, a closer look reveals considerable potential for technological modernization. With an overrepresentation of fossil energy sources, there are clear limits to a further reduction of energy demand and related GHG emissions. Thus, target group-specific measures aiming at technology change also appear reasonable in this group. At the moment, small impacts seem to be mostly a consequence of small-to-moderate usage intensity. However, this limited usage intensity appears not to be related to explicit personal intentions or pro-environmental self-identity but instead caused by other factors. Lifestyle 3 respondents show relatively often a low self-efficacy with regard to the energy transition, which might further reduce their motivation to change personal behavior or technology choices. On the one hand, labor market and social status give reason to assume that this group can afford technological upgrades more easily than other groups. On the other hand, the willingness to do so could be questionable.

Energy Lifestyle 4: "Homers"

Energy Lifestyle 4 shows an average overall energy demand of approximately 155,000 MJ per year. A total of 8.9% of all Austrian survey respondents were assigned to this group on the basis of their patterns of energy demand. Respondents from Vorarlberg/Tyrol and Styria are overrepresented in this group, while Upper Austria and Salzburg are underrepresented. Women and the ages 55+ are overrepresented, whereas 18–34 and 35–44 are underrepresented. There is a tendency toward smaller and single-person households without children. Large living space categories and rural housing types such as single-family homes are overrepresented. This, in combination with an elevated share of non-up-to-date heating technologies, results in the group having by far the highest average "housing"-related energy demand of all groups (320% of the Austrian average). Regarding education, the overall picture is diverse, and this group shows a tendency toward limited labor market integration. With regard to political orientation, center to center-right positions are overrepresented. There is a diverse picture of the subjective social status, with an elevated share of respondents who describe their situation as "worst off."

In "mobility," group members have a below-average energy demand. This reflects a high share of very small annual driving distances and a diverse overall picture with regard to air travel. There is some indication for openness toward more efficient car propulsion technologies (plug-in hybrid), whereas the share of hybrid-electric and fully electric vehicles is smaller than in the Austrian average. There is indication for little interest in car sharing, but also indication for openness toward public transport. Regarding bicycle use, this group shows very diverse behavior, with a tendency toward limited usage. The below-average energy demand for "consumption" reflects a tendency toward average or reduced consumption in fashion and a tendency toward reduced consumption in electronics. Despite a tendency toward reduced meat consumption, the overall picture of group-specific "diet"-related energy demand is again close to the average. Regarding "leisure" activities, there is an above-average share of activities that require a moderate amount of equipment and infrastructure, which results in the group having a below-average primary energy demand for "leisure." There is evidence for a limited use of electronics. Despite this, they have an above-average energy demand for "information," which is associated with the below-average share of respondents who use "green electricity."

In general, an intrinsic motivation seems to play only a minor role in potential upgrading decisions. Many respondents in this group strongly disagree with the intention to support the energy transition, and this group has an above-average share of respondents who do not believe in climate change. They often do not perceive any normative pressure to act in a way that would result in a lower energy demand, and they have low values in self-efficacy with regard to the energy transition. Concerning the possible creation of new jobs through the spread of renewable energy systems, strong disagreement is relatively often reported. Furthermore, this group shows a tendency toward little acceptance for costly policy measures. In view of the overall picture, financial incentives (especially for building-related upgrades) could be a starting point for a group-specific strategy.

Energy Lifestyle 5: "Consumers"

Energy Lifestyle 5 has an average overall energy demand of approximately 113,000 MJ per year. In total, 10.1% of all Austrian survey respondents were assigned to this group on the basis of their patterns of energy demand. Respondents from Burgenland and Styria are overrepresented in Energy Lifestyle 5, while Vienna/Lower Austria, Carinthia, and Salzburg are underrepresented. Women and ages 18–34 are overrepresented, whereas ages 35–54 are underrepresented. There is a tendency toward larger households. A high share of small living space and an elevated share of urban/suburban dwelling types are characteristic for this group. Although the picture regarding heating systems is diverse, the average "housing"-related energy demand of this group is below the national average. Regarding education, the overall picture is diverse, with some tendency toward higher education. Center-left political orientations are overrepresented. An above-average share of full-time students and respondents with good labor market integration is shown. Respondents reported a relatively high subjective social status.

Similar as respondents from lifestyle 3, respondents show a practically average overall energy impact, but they have a clearly distinct profile. In "mobility," this group is characterized by a slightly below-average energy demand. This reflects a high share of small-to-average annual driving distances and a diverse overall picture with regard to air travel. There is a tendency toward cars with higher efficiency propulsion systems. There is some indication for openness toward the use of public transport and some evidence for interest in and positive experience of car sharing. The group shows diverse behaviors regarding bicycle use but also an above-average share of all-season cyclists. The strongly above-average energy demand for "consumption" is a result of this group's tendency toward elevated or intensive consumption of fashion and electronics. Despite a meat-reduced overall picture of the group-specific "diet," the associated energy demand is again close to the average. Regarding "leisure" activities, there is an above-average share of activities that require a moderate amount of equipment and infrastructure, which results in this group having a "leisure"-related primary energy demand practically equal to the national average. The tendency toward an intensive use of electronics in combination with a below-average share of respondents using "green electricity" results in an above-average energy demand in the area of "information."

Lifestyle 5 has an above-average share of respondents who strongly agree with the statement that renewable energy systems are good for the environment, which matches with an overrepresentation of people who believe in climate change. Although many respondents reported to perceive normative pressure, it is not so common amongst members of this group to see it as a personal obligation to be energy efficient and to behave in accordance with the energy transition. Additionally, a moderate tendency to disagree with supporting the energy transition can be observed. However, there is also a slight overrepresentation of respondents who show higher self-efficacy with regard to the energy transition. In total, environmentally conscious consumption seems to already play some role in this group. This is, for example, supported by the overrepresentation of hybrid-electric and electric cars owned by respondents from this group.

6.5   Conclusions and Outlook

This chapter started from the assumption that an Energy Lifestyle is multidimensional. Using an energy-specific lifestyle method (Schwarzinger et al., 2019b), five Austrian Energy Lifestyles with distinct patterns in their energy-related behavior were identified. It turned out that there is no group with behavioral patterns that are even close to the "average Austrian." This is in line with findings on the European level, where no "average Europeans" could be identified in a relevant number. As found in an earlier study with an Austrian dataset from 2009 (Schwarzinger et al., 2018), "mobility," again, turned out to play a crucial role for Energy Lifestyles by showing the largest relative difference in energy demand between the groups. A similarly massive impact of "mobility" was found in a Europe-wide view (Schwarzinger et al., 2019a).

The five Energy Lifestyles identified in Austria were characterized in detail using the results of correspondence analyses. This reduction in complexity led to a relatively tangible picture of the Energy Lifestyle groups and their potential role in the energy transition. In most cases, group-specific behavior within and across different areas of life could be reasonably interpreted. Some examples for potential group-specific policy measures could be discussed for most groups, while, for example, in the case of lifestyle 2, the overall picture was less tangible, and a need for further research became evident. In general, this explorative methodology provided an overview with regard to the Austrian population that appears to be more useful than the typologies based on psychometric lifestyle models used in the past (Hierzinger et al., 2011; Bohunovsky et al., 2011).

Lifestyle-specific behavioral patterns, on the one hand, and sociodemographic and psychological characteristics, on the other hand, were put into relation with each other on the basis of the over- and underrepresentation of attributes in the individual groups. This approach takes into account that there is usually little statistical relationship between behavior and single explanatory variables when controlling for other variables (Kollmuss & Agyeman, 2002; Csutora, 2012; Newton & Meyer, 2013). The underlying causes for living a certain Energy Lifestyle can accordingly be manifold, and complex constellations of influencing factors might play a role. Using variables with an unreliable link to behavior as indicators for group membership might be a reason why earlier studies on Energy Lifestyles in Austria achieved less useful results and identified groups with relatively similar behavioral patterns.

The approach described here enhances the previous understanding of Energy Lifestyles by providing a reasonable and plausible overview of Energy Lifestyles in Austria. With regard to future research, it provides indications of where it could be worth taking a closer look in subsequent studies. The results presented show that an isolated view on single behavioral dimensions is at risk of not sufficiently considering the relationships between different behavioral domains. Furthermore, converting individual and group-specific behaviors into the "currency" of estimated primary energy demands (which is uncommon in sociology) makes the methodology a suitable basis for interdisciplinary cooperation, as has been demonstrated in the Horizon 2020 Project ECHOES.

When it comes to target areas for policy measures aiming at an effective reduction of energy demand and emissions on a national level, there is enormous potential for improvement in the energy-intensive areas of "mobility" and "housing." These sectors are those in which fossil fuels (and thus emission-intensive technologies) still play a major role. These priorities are hardly surprising and already part of strategies focusing on the technological transition toward zero-emission technologies. Despite the relatively strong focus on these two energy-intensive areas of life, the policy design for the five target groups poses the following very heterogeneous challenges:

In Energy Lifestyle 1, the group with the extraordinary high energy demand for "mobility," intrinsic factors do not seem to play a relevant role. Instead, linking the energy transition with the consumption of innovative technologies and related economic opportunities might be a more suitable strategy to engage this group in the energy transition. Energy Lifestyle 2, the largest group, which makes up almost half of the population, has a clearly lower average energy impact than the national average. However, this does not mean that this group should be disregarded in energy efficiency strategies. If the behavior and technological equipment of this group remain unchanged while the energy efficiency in the other groups increases, Energy Lifestyle 2 is likely to become problematic in terms of an "above-average" energy demand in the course of a few years. In the case of Energy Lifestyle 3, there is possibly an even greater risk that the group is considered unremarkable and overlooked due to its currently "average" energy demand. However, the group's impact is kept at this (from today's perspective) "average" level only by relatively low usage intensity. While the affordability of technological upgrades seems to be given, the motivation for environmentally sustainable consumption is low. The fact that respondents from 18 to 34 are overrepresented in this group draws special attention to group-specific long-term strategies. With regard to Energy Lifestyle 4, "Homers," potentials for energy savings might be accessed in particular through financial incentives for technological upgrades in the area of housing. However, an overrepresentation of single-person households and rural building types might be limiting factors for efficiency gains, posing a particular challenge to the design of energy strategies. Similar to respondents from Energy Lifestyle 3, "Hobbyists," respondents with Energy Lifestyle 5, "Consumers," have an average energy demand close to the Austrian national average. Remarkably, environmentally conscious consumption and practices, especially in the area of "mobility," already play a certain role for this group and might be taken into consideration for specific policy strategies.

Of course, these policy potentials only represent a small fraction of the considerations that need to be taken into account in the preparation of nation-wide energy efficiency and decarbonization strategies. However, with the exploration of five Energy Lifestyles in Austria, a basis for a better understanding of affected target groups' life realities could be achieved. The better an individual's specific needs can be taken into account, the more likely it is to improve their engagement in the energy transition, and the higher the chance is to reconcile energy efficiency, social sustainability, public acceptance, and economic feasibility. The next chapter thus will consider individuals' obstacles to change and the gap between their attitudes and behaviors.

References

Bird, D. N., Schwarzinger, S., Kortschak, D., Strohmaier, M., & Lettmayer, G. (2019). *Report: A detailed methodology for the calculation of cumulative energy demand per survey respondent.* The ECHOES Consortium.

Bohunovsky, L., Grünberger, S., Frühmann, J., & Hinterberger, F. (2011). *Energieverbrauchsstile Datenbank zum Energieverbrauch österreichischer Haushalte: Erstellung und empirische Überprüfung.* Endbericht.

Bourdieu, P. (1987). *Distinction: A social critique of the judgement of taste.* University Press.

Csutora, M. (2012). One more awareness gap? The behaviour–impact gap problem. *Journal of Consumer Policy, 35,* 145–163.

Finkbeiner, M., Inaba, A., Tan, R., Christiansen, K., & Klüppel, H.-J. (2006). The new international standards for life cycle assessment: ISO 14040 and ISO 14044. *The International Journal of Life Cycle Assessment, 11,* 80–85.

Hierzinger, R., Herry, M., Seisser, O., Steinacher, I., & Wolf-Eberl, S. (2011). *Energy Styles. Klimagerechtes Leben der Zukunft—Energy Styles als Ansatzpunkt für effiziente Policy Interventions. Endbericht zum Projekt Energy Styles.* Klima- und Energiefonds.

Kollmuss, A., & Agyeman, J. (2002). Mind the gap: Why do people act environmentally and what are the barriers to pro-environmental behavior? *Environmental Education Research.* Routledge, *8,* 239–260.

Lüdtke, H. (1996). Methodenprobleme der Lebensstilforschung. Probleme des Vergleichs empirischer Lebensstiltypologien und der Identifikation von Stilpionieren. In O. G. Schwenk (Ed.), *Lebensstil zwischen Sozialstrukturanalyse und Kulturwissenschaft* (pp. 139–163). VS Verlag für Sozialwissenschaften.

Newton, P., & Meyer, D. (2013). Exploring the attitudes-action gap in household resource consumption: Does "environmental lifestyle" segmentation align with consumer behaviour? *Sustainability, 5,* 1211–1233.

Reichl, J., Cohen, J., Kollmann, A., Azarova, V., Klöckner, C., Royrvik, J., Vesely, S., Carrus, G., Panno, A., Tiberio, L., Fritsche, I., Masson, T., Chokrai, P., Lettmayer, G., Schwarzinger, S., & Bird, N. (2019). *International survey of the ECHOES project.* Zenodo.

Schwarzinger, S., Bird, D. N., & Hadler, M. (2018). The "Paris lifestyle"—Bridging the gap between science and communication by analysing and quantifying the role of target groups for climate change mitigation and adaptation: An interdisciplinary approach. In W. Leal Filho, B. Lackner, & H. McGhie (Eds.), *Addressing the challenges in communicating climate change across various audiences* (pp. 375–397). Springer International Publishing.

Schwarzinger, S., Bird, D. N., Lettmayer, G., Henriksen, I. M., Skjølsvold, T. M., Olaeta, X. U., Alvarez, L. P., Velte, D., Iturriza, I. J., Biresselioglu, M. E., Demir, M. H., Dimitrova, E., Tasheva, M., Tiberio, L., Panno, A., Carrus, G., & Costa, S. (2019a). *Comparative assessment of European energy lifestyles.* ECHOES Project Consortium.

Schwarzinger, S., Bird, D. N., & Skjølsvold, T. M. (2019b). Identifying consumer lifestyles through their energy impacts: Transforming social science data into policy-relevant group-level knowledge. *Sustainability, 11.*

CHAPTER 7

Obstacles to Lower Environmental Impact in Low-Cost Behaviors

The previous chapters focused on measuring the overall greenhouse gas (GHG) output, the factors that shape an individual's total emissions and different patterns of energy consumption. This chapter[1] turns toward the gap between environmental values and behaviors as well as the obstacles in lowering one's environmental impact. Special emphasis is placed on "low cost behaviors" in the areas of mobility and consumption.

A gap between attitudes and behaviors is quite common. The Eurobarometer 2008, for example, reports that a total of 96% of respondents from all European countries consider environmental protection to be of great importance (64% very important and 32% quite important) and that 75% say they are "willing to buy environmentally friendly products even if they are a bit more expensive" (Spezial Eurobarometer, 2008, p. 12 and pp. 29ff). However, only 17% of the respondents actually bought more eco-friendly products despite their higher price in the month before the survey. This chapter thus looks into the question of which obstacles impede consistent environmentally conscious action. This question will be addressed using a mixed-methods approach, exemplified by two environmentally relevant behaviors, consumption, and mobility. Finally, guidelines for action and policy measures to promote environmentally friendly behavior will be proposed.

---

[1] Lead author: Beate Klösch. This chapter is a continuation of the lead author's MA thesis.

© The Author(s) 2022
M. Hadler et al., *Surveying Climate-Relevant Behavior*,
https://doi.org/10.1007/978-3-030-85796-7_7

7.1   Theoretical Approaches to Inconsistencies in Environmental Behavior

An inconsistency between a person's attitude and behavior is known in the scientific discourse (Blake, 1999; Brown & Sovacool, 2018) as the value-action gap (or attitude-behavior gap). Scientific findings indicate a demonstrable discrepancy between personal attitudes and actual behavior, especially regarding environmental actions. As early as 1999, Blake applied the concept of the value-action gap within an environmental context. Since then, this discrepancy has been proven in numerous studies on different behaviors (such as avoidance of generating refuse, recycling behavior, or the purchase of environmentally friendly vehicles) and different countries (Barr, 2004, 2006; Chung & Leung, 2007; Hadler et al., 2019). A number of studies can also be found regarding the reasons for the occurrence of the environmental value-action gap (see Blake, 1999; Kollmuss & Agyeman, 2002; Neugebauer, 2004; Mairesse et al., 2012; Chaplin & Wyton, 2014). Blake (1999) identifies individual barriers, a sense of responsibility and practical feasibility as primary obstacles to pro-environmental behavior. Kollmuss and Agyeman (2002) also distinguish similar barriers, including internal factors (motivation, environmental knowledge, values, and responsibility), external factors (institutional, economic, social, and cultural factors), and socio-demographic factors (gender and number of years of education). As an explanation for the occurrence of a value-action gap, Neugebauer (2004) also mentions internal conflicts of objective caused by the presence of competing behavior-relevant attitudes of a person as well as habits or stress.

From the literature reviewed, three main insights emerge. First, environmental behavior and the value-action gap are complex phenomena and depend on a variety of factors. Second, in several studies, intention proved to be a particularly meaningful predictor of environmental behavior, in some cases even more valid than attitudes (Barr, 2004, 2006; Ajzen, 2012). Third, most of the studies reviewed refer to the Theory of Planned Behavior (Ajzen, 1985), mentioned in Chap. 2. As it is one of the most widely used theories to explain environmental behavior and reflects the central factors of this analysis, it is also used as a theoretical framework in this chapter. This concept attempts to predict behavior using the parameters of attitude, norm, behavioral control, and intention. A central aspect is that attitudes, norms, and perceived behavioral control do not directly determine a person's behavior, but rather first influence personal intention, which ultimately shapes our behavior (Ajzen, 2012; Chao, 2012). If one or more of the three initial elements (attitude, subjective norm, and behavioral control) are not congruent with the intention, or the intention is not congruent with the actual behavior, a gap arises between these factors. In previous studies, however, the focus has primarily been on attitude and behavior, while little attention has been paid to intention. Therefore, intention will be included in this work, making value-intention or intention-action gaps possible.

The Framework of Environmental Behavior (Barr, 2004), which comprises a similar structure to that of environmental behavior, is further considered to complement the Theory of Planned Behavior. Barr additionally incorporates situational factors (social context, socio-demographics, knowledge, experiences, and possibilities of execution) as well as psychological factors (motivation, subj. Norms, and self-efficacy) (see Barr, 2004, pp. 234 f.). Based on these two theoretical concepts, a simplified linear model is used in this chapter, which depicts all the variables examined in the first empirical step. It is assumed that personal attitudes toward environmental problems and environmentally relevant behavior influence one's intention, which in turn shapes the behavior actually carried out. However, it must be anticipated that there may be deviations in the behavioral prediction and thus gaps between attitude, intention, and behavior. Since it is as of yet unclear whether these discrepancies occur between all three variables (i.e., between either attitude and behavior, attitude and intention, or intention and behavior) or whether the gap can only be observed between two of them, the theoretical model remains openly framed.

Finally, the low-cost hypothesis of environmental behavior (Diekmann & Preisendörfer, 1998), which was also presented in Chap. 2, is used to determine the behaviors to be studied. It describes humans as rational actors, the homo oeconomicus, who make decisions based on cost-benefit assessments. Accordingly, the importance of one's own environmental attitudes decreases, even among particularly eco-conscious individuals, when effort or costs rise (Diekmann & Preisendörfer, 2001; Preisendörfer, 1999). The focus of this chapter is on two behavioral dimensions with high environmental impact—mobility (kilometers driven by car per year) and consumption (new purchase of particularly CO_2-intensive products). These behaviors were chosen because of their low-cost character, and research thus would predict a strong congruence. Secondly, these two behaviors are particularly CO_2 intensive. By analyzing these two behavioral dimensions, both of which can be carried out and changed by the individual without high costs, comparisons between similar environmental behaviors become possible. In addition, fields of action open up to promote environmentally friendly low-cost and high-impact behavior.

7.2   Identifying Value-Intention-Action Gaps in Our Sample

The current analysis of value-intention-action gaps uses the dataset described in detail in Chaps. 4 and 5. The research questions focus on whether and which of the aforementioned gaps can be detected and for which subjects they are more likely to occur. Environmentally relevant gaps are defined as the discrepancy between attitudes or intentions and actual behavior. In Table 7.1, such a gap is exemplified by the intention-action gap in mobility and consumption. According to this, about 15% of the respondents have a high intention to act in an environmentally friendly way but still travel many kilometers by car per year (median split). As for consumption, over 22% of the survey respondents act against their high environmental intention and show a high consumption of CO_2-intensive products.

All three types of initially possible gaps[2] are present in our data (in 10–35% of the sample). The centerpiece of this study, however, is the gap between a "high reduction intention" and a "high usage," which is highlighted in Table 7.1. Respondents show value-action and intention-action gaps more frequently in consumption than in mobility. According to this, individuals may find it more difficult to implement their environmentally positive attitudes and intentions in their consumption behavior than in their mobility behavior although both behaviors concern low-cost situations that could be shaped quite easily by actors.

Based on multiple regression analyses, some significant differences depending on socio-demographic characteristics, especially regarding

---

[2] Given that the regression analysis for the value-intention gap was not significant, it seems that there are no socio-demographic differences in the probability of occurrence of a gap between attitude and intention for the overall population. Thus, it appears that positive environmental attitudes translate equally into correspondingly consistent intentions to protect the environment. Accordingly, consistency does not seem to fail at the transition from attitude to intention, and individuals intend to change their behavior based on their environmental attitudes. Only when transferred into actual behavior do group-specific differences seem to occur.

Table 7.1 Intention-action gap in mobility and consumption

Actual car use	Intention to change		Actual consumption	Intention to change	
	Low	*High*		*Low*	*High*
Low	31.6%	33.7%	Low	23.3%	26.7%
High	19.9%	14.8%	High	27.7%	22.3%

N = 196
Source: OeNB sample (Hadler et al., 2021)

value-action and intention-action gaps, were identified. Age, residential area, and household constellation proved to be particularly influential. Put together, younger respondents, individuals living in rural areas, and families with children under 18 and high income are more likely to show a discrepancy between their attitudes or intentions and their behavior. Therefore, the focus of this subsequent qualitative study is on the obstacles that inhibit the environmentally responsible behavior of these groups.

7.3 An In-Depth Look at Obstacles to Environmentally Friendly Behavior and Solutions

"At first sight, the nature of a value-action gap suggests either hypocrisy or non-understanding, however […] the situation is more complex" (Chaplin & Wyton, 2014, p. 204). The central question thus is to identify the reasons for the discrepancies between attitude, intention, and behavior, especially among individuals of different age, place of residence, and household constellation. The following sections present the results from qualitative semi-structured interviews, with participants selected based on the quantitative results presented in the previous section.

The following two conditions were found to be important for recruiting the sample: the respondents should have a positive attitude toward environmental issues, and they should own a car or at least use one regularly. Furthermore, an even distribution of the relevant socio-demographic variables (residential area, age, and household type) was aimed for: A total of 15 interviews were conducted with 16 respondents (one of which was with a couple); seven of the interviewees are women and nine live in rural areas. Furthermore, an age distribution from 23 to 68 years was achieved. For later analyses, the sample was divided into younger (23–36 years, 7 in total) and older (49–68 years, 9 in total) respondents. Regarding household constellations, the sample was divided into three groups, as follows: four single households, four family households with children, and seven multi-person households. Multi-person households mainly include couples as well as shared apartments with friends and living with a parent. For the following analyses, the household form is categorized in two different ways—first, depending on the presence of children (4 persons with children, 11 without), and secondly, depending on the number of people (5 singles, 10 partnerships/shared flats/with a parent).[3]

Following Mayring's method of qualitative content analysis, the interviews were inductively coded and analyzed. Due to the large amount of data, the focus is placed on the following three areas:

1. Obstacles to environmentally friendly behavior: Why do individuals not act in an environmentally friendly way but rather contrary to their eco-friendly attitudes?
2. Individual requirements: What would the interviewees wish for/ what would be necessary to make it easier for them to act in an environmentally friendly way in their daily lives?
3. Social strategies: How would people change their actions toward a more environmentally friendly way?

These three questions come into focus as they all explore the cause of value-intention-action gaps and provide suggestions for solutions.

7.4   Obstacles to Environmentally Friendly Behavior

First, obstacles to environmentally friendly action mentioned in the interviews were analyzed and coded. A total of 16 codes resulted, which can be categorized into two dimensions of structural and intrapersonal factors. Structural conditions mainly include situational factors, such as the limited availability of goods and services, lack of information, and time and cost factors. Intrapersonal factors are obstacles at the individual level, such as convenience, routine, or lack of interest.

---

[3] It should be noted that due to the uneven distribution of group sizes regarding household constellation (both in terms of the number of people living in the household and children), the interpretability of the results is limited.

Time and cost factors of environmentally friendly alternatives prove to be the most influential structural framework conditions. Especially high prices for public transport and higher pricing of sustainable clothing or biological food, compared to conventional products, were often emphasized. The time factor frequently refers to public transport, which, on the one hand, often requires a long wait, and, on the other hand, usually takes longer than if one were to travel by car. Higher costs of public transport were also often mentioned. In combination with the time required, it was also mentioned that eco-friendly actions are often cumbersome (be it in terms of consumption, when you have to visit ten shops to find all products of organic quality, or be it concerning mobility, where the use of public transport with children, heavy luggage, etc., is too much of a hassle). The fundamental lack of environmentally friendly products on offer, such as in clothing, was also a frequently cited obstacle for not acting in an environmentally friendly way.

To conclude, for many of the interviewees, the availability of eco-friendly alternatives such as clothing and public transport is not sufficient, and the existing supply is often expensive or involves a lot of effort. Another obstacle is the lack of information or knowledge about more environmentally conscious alternatives. All in all, information is lacking at several points. First, there must be information or knowledge about which behaviors are harmful to the environment and that there are alternatives. Secondly, there is often a lack of information on what these alternatives look like. And third, there is a lack of information on where these alternative products (e.g., in regional or sustainable quality) are available. This shows that despite the individual willingness of the interviewees to inform themselves, there is a lack of clear and public information that can contribute to a more environmentally friendly behavior. Another factor that was mentioned several times is the credibility of eco-friendly products. Several interview partners reported that they are often not sure which is the better alternative or that they have limited trust in certifications, labels, and the like. One interviewee commented on this as follows:

I sometimes have the feeling that just because it says organic somewhere, it doesn't necessarily have to be organic in the classical sense, or just because it says sustainable products or sustainable materials [...] So I still have the feeling that loopholes are found or made, and people just take it along or offer it, because that's the way it is now. [...] So for me it would be very important that people know that they can rely on it if it says so.

The most influential intrapersonal obstacle to environmentally friendly behavior is the convenience of those interviewed. This affects both the search for information on eco-friendly offers and actually carrying out the behavior. This was often explained by the attempt to avoid undue effort (e.g., doing the shopping by public transport). Routine and habit were also frequently cited as reasons why interviewees did not make use of environmentally friendly options. In addition, the lack of consistency in one's own actions was mentioned a few times, and environmentally harmful behavior, especially in the area of consumption, was justified by frugality. This applies in particular to the purchase of new clothing by men.

Some links between structural and intrapersonal factors also emerge. In particular, the time factor was often mentioned together with personal comfort level and the inconvenience of environmentally friendly alternatives as a barrier to eco-friendly behavior. There also seems to be a connection between the cost factor and convenience, similar to the lack of supply. Routine also frequently occurs together with the time factor as well as the lack of supply.

A comparison of the barriers for the two behaviors, mobility and consumption, reveals some obstacles that occur in both. Commonalities occur regarding time and cost factors, convenience, and lack of supply. However, the time factor and convenience act as barriers to environmentally friendly behavior more often in the context of mobility than in consumption. Factors that have no effect on mobility behavior but that inhibit environmentally friendly consumption behavior are especially lack of information, no clear labeling of eco-friendly products or doubts in the credibility of existing labels as well as a lack of consistency in using environmentally friendly products.

Subsequently, the reasons for the occurrence of value-intention-action gaps were analyzed in relation to influential socio-demographic characteristics, specifically, place of residence, age, and household structure. With regard to place of residence, there is definitely a difference in obstacles to environmentally friendly behavior between respondents living in rural areas and those living in the city. The dependence on place of residence was mentioned several times by interviewees in rural areas, whereas it was not mentioned by urban dwellers. The statements in this regard vary from a lack of public transport connections, a smaller range of stores/options, and so on to financial advantages since regional fruits and vegetables can be purchased more cheaply at the rural farmers' market than in the city. Nevertheless, it is evident that individuals living in rural areas are more aware of their location's dependency and the resulting barriers than those living in the city. Cost and time factors were also mentioned more frequently in rural areas than in the city, as were high costs and lack of options (especially when it comes to clothing) and information as well as individual lack of consistency. In comparison, respondents in the city more often mentioned their habits and routines as a reason for a lack of environmental behavior.

There is one barrier that was only mentioned by older interviewees, namely forgetting or not thinking about more environmentally friendly alternatives. An older interviewee said,

> *I have difficulties remembering to take a box with me when I go shopping, like it was done in the 80s. We did all that before, and somehow it completely dropped off in the 90s. And no one went shopping with their own dishes anymore. And to think of that again.*

Furthermore, older individuals mentioned more often the lack of information and the cost factor than younger ones. The latter seem to see greater problems in the time factor (especially concerning public transport) and their convenience.

There are also some differences between households. Comparing single- and multi-person households shows that forgetting or not thinking about eco-friendly alternatives was only mentioned in multi-person households. In addition, respondents living with more than one person also mentioned convenience, cost, and time factors, as well as lack of information as reasons for their partial lack of pro-environmental actions. Furthermore, some reasons were only mentioned by respondents without children. These were general structural conditions and the lack of labeling of environmentally friendly products. On the intrapersonal level, respondents without children mentioned a lack of interest in learning more about eco-friendly options as well as frugality in certain areas. Surprisingly, the interviewees with children in their household did not mention any reasons that are not equally common in childless households.

Overall, a complex interplay of numerous reasons for the emergence of value-intention-action gaps can be identified. A comprehensive intervention program seems to be needed that promotes environmentally friendly behavior in all areas, both on the individual and on the societal levels. This view is further examined in the following sections based on the statements of the interviewees.

7.5   Individual Requirements to Facilitate Environmentally Friendly Behavior

After knowing why it is often not possible for the interviewees to choose environmentally friendly alternatives and behaviors, this section looks into what they would need or wish for to make it easier to act in an environmentally friendly way.

The respondents most frequently wished for more information on environmentally friendly products and a stronger presence of environmental issues in the public discourse. Information on what is actually eco-friendly and what options and offers are available is desired. According to the interviews, the mobility sector also seems to be particularly expandable. The interviewees frequently wished for better public transport, which concerns the geographical expansion of these as well as more frequent connections (including evenings and weekends). Public transport should also be cheaper, and reference was made to pay structures such as in Vienna (€1 per day). Alternatives were also mentioned, such as a planned regional train project or e-car sharing offered by the municipality. In the area of consumption, the interviewees often wished for better labeling of environmentally friendly products, such as through a unified seal or through signs in shops. This wish was expressed several times for both clothing and food, and some suggestions for implementation came from the respondents themselves (labeling by, e.g., green tick, environmental calories, and/or CO_2 footprint on each product). One respondent expressed the following:

> Maybe on top of packaging, it would be quite a good idea, because everyone knows the table with calories and nutritional value, but maybe also such an ecological table that you just know how many liters of water were used for the production of that product, from the beginning to the end. Simply how many liters of water were used, how many tons of CO_2 were emitted, for that piece you have there.

There also seems to be a big deficit in the availability of environmentally friendly products in both rural and urban areas. The respondents would like to see a larger and regional range of food (preferably a shop where you can get everything at once), more packaging-free shops and a larger range of sustainably produced clothing. There appears to be a backlog here, especially for men's clothing and fashionable items. Several respondents would also like to see better support for environmentally friendly projects, be it independent fashion shops or so-called environmental banks. One interviewee talked about such a bank that only gives loans to regional organic farmers. Others would also like to see more financial support from the municipality for local shops.

Finally, individual requirements to facilitate eco-friendly behavior were also considered with respect to dependencies on the socio-demographic characteristics of place of residence, age, and household constellation. Regarding the place of residence, it can be shown that only respondents living in rural areas would like to have reduced public transport fares and other cheaper alternatives in the context of consumption. Thus, the cost factor seems to play a greater role in rural areas than in cities. With regard to age, there are only minor differences among individual requirements. There is a tendency that younger individuals would like to see a wider range of clothing and other environmentally friendly products and their labeling as well as a better range of public transport. Concerning the household type, it appears that respondents from a multi-person household would like better public transport as well as a wider range of eco-friendly products such as for food. These factors were not mentioned in the single households. Furthermore, the former group tends to require more information on and the presence of the topic in the public discourse, better labeling of products, funding of environmentally friendly projects, and a wider range of clothing and public transport, as well as cheaper public transport fares than do respondents living on their own. Likewise, respondents without children in the household would like to see cheaper alternatives, which were not mentioned by respondents with children in the household. Otherwise, respondents without children tended to wish for more information on and the presence of the topic in daily life and a broader offer with better labeling as well as discounts on public transport.

7.6 Social Strategies to Promote Environmentally Friendly Behavior

After discussing subjective obstacles to environmentally friendly behavior and individual requirements to facilitate the implementation of eco-friendly alternatives, this section takes a brief look at what is needed, according to the interviewees, to achieve a change in behavior in society as a whole. The interview participants most frequently saw potential for social change among the population through education and raising awareness about the climate crisis. The problem of lack of information was also mentioned several times:

> *I think for many people the problem is often the beginning because they don't know how or where. How can I change something, where ...? And there we come to the next issue, the flow of information. That somehow you would have to teach or explain that to people like that ... the education.*

According to the interviewees, more information and education should be provided through the media, through targeted advertising and famous personalities acting as role models (such as Greta Thunberg, Arnold Schwarzenegger, influential politicians, or athletes). Equally important and necessary for the interviewees is the introduction of regulations and prohibitions by policymakers, both for individuals and for large players, such as companies. One interviewee formulated this point as follows:

> *Unfortunately, I think it is important for the majority to have bans. Yes, personal freedom is also important, but I think that when it comes to such a topic that simply affects us all and where the effects are so massive, and where science has been warning for a long time and somehow, we don't manage to change behavior simply because of a raise in awareness, then I have the feeling that you have to force people to do it.*

Outside the individual level, guidelines for companies and policies regarding the transportation sector as well as legal regulations, such as for packaging, are desired. Sanctions should be enforced in cases of non-compliance with the given laws. This includes the introduction of a CO_2 tax or kerosene tax for air traffic. The interviewees see another influential factor in schooling—the necessity of implementing environmental topics and knowledge in schools as well as the discussion of consequences and alternatives was mentioned several times, as children would in turn pass this awareness on to their families. Finally, some respondents also consider a systemic change to be necessary in order to counter the climate crisis. Here, a change in values away from current mass consumption is emphasized. According to the respondents, a rethinking of what quality of life means is needed, including in the context of work and leisure. Other solutions that were brought up repeatedly are, on the one hand, a development toward more regionality and the need for environmentally conscious lifestyles without any feelings of restriction. Such normality should be conveyed through school and should find its way into the personal sphere, especially the family, and become part of the habitus. It was also mentioned a few times that an environmentally friendly lifestyle must become more attractive. It should become financially as well as personally attractive, stand for a better quality of life, and be considered a status symbol. Finally, it can be concluded that the interviewees see the greatest potential in political guidelines and bans as these offer the opportunity to exert influence on daily actions of both individuals and larger entities such as companies on many different levels.

7.7 DISCUSSION OF OBSTACLES TO ENVIRONMENTALLY FRIENDLY BEHAVIOR AND POTENTIAL SOLUTIONS

In summary, the analysis of the interviews shows that time and cost factors (structural reasons), as well as convenience and routine (intrapersonal reasons), were the major obstacles to consistency between attitude, intention, and behavior in regard to environmentally friendly behavior. This finding also coincides with previous results from the literature, such as with the individual and practical barriers found by Blake (1999). However, taking responsibility and perceiving one's own actions as effective do not seem to be a problem for the interviewees in this study. Almost all agreed that the contribution of individuals (in addition to political framework conditions) is crucial to counteract the climate crisis. Likewise, Kollmuss and Agyeman's (2002) "Model of Pro-Environmental Behavior" includes some factors that were confirmed here as barriers to pro-environmental action. In particular, the lack of external opportunities such as infrastructure, economic situation, and political conditions, as well as old habits of action, seems to have negative effects in the sample. The internal factors predicted by Kollmuss and Agyeman, such as personality traits or lack of environmental awareness, appear to be less substantial in this study.

The qualitative approach allows us to delve deeper into the (dis)similarities between environmentally relevant mobility and consumption behaviors. In this respect, it can be concluded that, contrary to the theoretical assumption, discrepancies between attitude, intention, and behavior can also occur in environmentally related low-cost behaviors, which can be explained to a large extent by the obstacles found in the present analysis. The respondents face the same barriers in both behaviors (e.g., time and cost reasons and lack of supplies) but also behavior-specific obstacles such as the lack of labeling of eco-friendly products. For the mobility realm, factors of time, cost, and convenience were particularly inhibiting. These results are in line with the findings of Diekmann and Preisendörfer (see Diekmann & Preisendörfer, 2001, p. 73) as well as with the assumption of the rational choice theory that these very factors have the greatest influence on behavior, for example, in the choice of means of transport, especially in the context of mobility (see Götz, 2011, p. 334). The results of this study suggest that it would often be possible to switch to public transport but that this is often not done due to the reasons mentioned above. This problem could potentially be addressed with appropriate (e.g., financial) incentives and motivators to reduce the costs of engaging in eco-friendly behavior.

In comparison, considerably more obstacles were found regarding consumption behavior. The large number of possible barriers in consumption could explain why, according to the quantitative results, discrepancies occur more frequently here than in the mobility realm. In the context of consumption, eco-friendly behavior could be supported in particular by improving the structural conditions in terms of supply and labeling. It seems that personal environmental attitudes and existing intentions are often not enough to encourage environmentally friendly behavior even if the situations are theoretically low-cost and individuals could easily adapt.

Furthermore, respondents in rural areas are confronted with different obstacles than those living in cities, especially considering the availability of alternatives to private cars, clothing, and other eco-friendly products. The cost factor is also a greater obstacle in rural areas than in cities. This may be due to different reasons. On the one hand, environmentally friendly alternatives in terms of mobility and consumption may actually be cheaper in cities as there is more choice there. It can be assumed that this is especially the case for public transport, which is cheaper as well as better developed in cities. On the other hand, this perception could possibly also be due to the sample and the distribution of socio-demographic variables such as age and income (the individuals living in cities tended to be younger; no data are available on income). Nevertheless, it should be noted that due to location dependency, the precondition that actions must be objectively possible is often not met in rural areas (see Tanner & Foppa, 1996, p. 246). It can be concluded that a low-cost behavior can become a high-cost behavior depending on the place of residence. Finally, it is the individual's "definition of the situation" that matters; as Diekmann and Preisendörfer sum up: it is about the perception of alternative actions, the assessment of the occurrence of consequences of actions and the perception of costs and benefits of the consequences of actions (see Diekmann & Preisendörfer, 2001, p. 76). Accordingly, environmentally friendly behavior is often more cost-intensive in rural areas than in cities, which subsequently leads to a more frequent occurrence of discrepancies between attitude, intention, and behavior in rural areas.

Moreover, older respondents reported that they often do not think about or forget about more environmentally friendly alternatives. One explanation for this could be the decades of routine in everyday activities such as grocery shopping. Regarding the household constellation, interviewees from multi-person households reported problems with convenience and lack of information as well as time and cost factors. One could conclude here that single households have more money and time available to choose more eco-friendly alternatives. Another explanation could be the increased obligations that additionally influence individuals in multi-person households in their decisions to act. However, respondents with children do not seem to face any additional obstacles. It can be concluded that multiple obligations, such as raising children, do not provide additional obstacles to environmentally friendly behavior that do not also occur in persons without children. Thus, the results of the quantitative study that people with children more often have a value-intention-action gap cannot be adequately explained through these interviews.

In summary, a large number of barriers to environmentally friendly action was found both on the structural and intrapersonal levels. The results show that, often, several obstacles exist at the same time when there is a discrepancy between one's attitudes, intentions, and behavior. Accordingly, two birds could be killed with one stone here, for example, by counteracting the intrapersonal factors such as convenience and routine by creating attractive offers and a better flow of information on the structural level.

Finally, some individual needs and societal solutions could be worked out. On the individual level, better availability of public transport and a wider choice of eco-friendly alternative products as well as their unified and clear labeling were often considered necessary. The desire for the normality or self-evidence of environmentally friendly products and behaviors as well as an accompanying reduction of the offer of environmentally harmful options was also mentioned repeatedly. Thus, there seems to be a need for action, especially at the structural level, which was also reflected in the social strategies and potential solutions proposed by the interviewees. To induce a change in behavior in society, the interviewees see a necessity in political measures, regulations, and bans. These regulations should be enforced with the help of sanctions or a corresponding CO_2 tax. Equally relevant to the interviewees seems to be a stronger education of the population to create greater awareness of nature and its protection. This should begin in early education and, if necessary, also be promoted on an institutionalized level in the form of environmental school lessons. In this context, the importance of eco-positive role models was also emphasized, through which young people can orientate themselves. Following this, on both the individual and societal levels, it was frequently mentioned that more information is needed to strengthen environmentally friendly action and to circumvent any related obstacles. This was described by two interviewees as follows:

> I thought it would be cool if it was somehow explained in detail on their homepage or if there was a video about what exactly needs how many resources. Just more information. So, for me, it always helps quite a lot if I know something about it, so that I then decide in favor of the alternative, even if it is perhaps inconvenient.
>
> Well, it's shocking, but if you publicize the whole thing, people will certainly think about it. And science is particularly important in my eyes, that they show the whole thing, what effects it has: Because then you will certainly be able to reach more people.

However, scientific discourse and the dissemination of information on climate change were found to be ineffective in changing opinions. Rather, opinions were reinforced by the information communicated due to selective information processing. According to Moser and Berzonsky, a "louder one-way messaging will only add to polarization rather than reduce it" (Moser & Berzonsky, 2015, p. 17), which is why they emphasize the need for individual motivation and willingness to act for the environment. Thus, other potential solutions should be considered more closely.

With regard to the differentiation between attitude and intention, no clear findings can be derived from the interviews. In general, it can be assumed that most of the interviewees have both positive environmental attitudes and the willingness to change their own behavior for the environment. According to the statements, this often fails due to a lack of options for action and offers, which ultimately lead to an intention-action gap. Only a few respondents stated that they have no interest in specific topics and thus no intention to change their behavior. However, this only concerns a few older men when they were asked about their consumption behavior in terms of clothing. They also justify their lack of interest through their lower consumption of clothing. Thus, the value-intention gap, which already played only a minor role in the quantitative analyses, is also to be expected here only to a very small extent and under certain circumstances.

7.8   Conclusions and Outlook

The aim of this chapter was to understand why environmentally conscious individuals nevertheless fail to behave in an environmentally friendly way. The quantitative analysis showed that individuals find it more difficult to behave consistently with their attitudes and intentions in the context of consumption than in that of mobility. The qualitative interviews highlight that individuals are most often deterred from the desired behavioral practice by intrapersonal factors (such as their own convenience or habits) and structural conditions (such as time and cost factors, lack of offerings).

These findings are mostly in line with the results of other studies. Regarding the obstacles to eco-friendly behavior, many elements from the "Model of Pro-Environmental Behavior" by Kollmuss and Agyeman were confirmed, specifically, the importance of structural conditions as well as routines and habits for pro-environmental behavior. The importance of situational factors is also in line with Barr's "Framework of Environmental Behavior."

As for the question of how to promote environmentally positive behavior in society, the interviewees considered two steps to be particularly necessary—on the one hand, there is a need for increased education, information dissemination, and awareness raising among the population in order to counteract intrapersonal barriers. These can be promoted through implementation in school lessons or via media and advertising, which should eventually lead to a change in lifestyle, away from consumption, as well as a normality of environmentally friendly behavior.

On the other hand, at the structural and political levels, a change in market regulation is desired, especially with regard to the offering and promotion of eco-friendly alternatives. Likewise, a restriction on environmentally harmful products and behaviors should be aimed for, both through bans and associated sanctions as well as CO_2-related taxes. Hence, a whole range of design proposals can be derived from the interviews, such as one unified and credible label for pro-environmental products, an indication of environmental or "CO_2 calories" depending on the CO_2 emissions for each product, and a CO_2 app that documents daily consumption and introduces alternatives in a playful way.

There were also numerous calls for action in the interviews, which should be heard above all at the political level. Although each of the interviewees considered the individual's contribution to environmental protection to be essential, there was consensus that policies need to provide structural conditions in order to exploit the full potential. The concluding chapter will thus consider some of these suggestions in an international comparative analysis and test their effectiveness.

References

Ajzen, I. (1985). From intentions to actions: A theory of planned behavior. In J. Kuhl & J. Beckmann (Eds.), *Action control: From cognition to behavior* (pp. 11–39). Springer.

Ajzen, I. (2012). The theory of planned behavior. In P. A. M. Lange, A. W. Kruglanski, & E. T. Higgins (Eds.), *Handbook of theories of social psychology* (pp. 438–459). Sage.

Barr, S. (2004). Are we all environmentalists now? Rhetoric and reality in environmental action. *Geoforum, 35*, 231–249.

Barr, S. (2006). Environmental action in the home: Investigating the 'value-action' gap. *Geography, 91*(1), 43–54.

Blake, J. (1999). Overcoming the 'value-action gap' in environmental policy: Tensions between national policy and local experience. *Local Environment, 4*(3), 257–278.

Brown, M. A., & Sovacool, B. K. (2018). Theorizing the Behavioral dimension of energy consumption: Energy efficiency and the value-action gap. In D. J. Davidson & M. Gross (Eds.), *The Oxford handbook of energy and society* (pp. 201–221). Oxford University Press.

Chao, Y.-L. (2012). Predicting people's environmental behavior: Theory of planned behavior and model of responsible environmental behavior. *Environmental Education Research, 18*(4), 437–461.

Chaplin, G., & Wyton, P. (2014). Student engagement with sustainability: Understanding the value-action gap. *International Journal of Sustainability in Higher Education, 15*(4), 404–417.

Chung, S., & Leung, M. M. (2007). The value-action gap in waste recycling: The case of undergraduates in Hong Kong. *Environmental Management, 40*(4), 603–612.

Diekmann, A., & Preisendörfer, P. (1998). Umweltbewußtsein und Umweltverhalten in Low-und High-Cost-Situationen: Eine empirische Überprüfung der Low-Cost-Hypothese. *Zeitschrift für Soziologie, 27*(6), 438–453.

Diekmann, A., & Preisendörfer, P. (2001). *Umweltsoziologie. Eine Einführung.* Rowohlt.

Götz, K. (2011). Nachhaltige Mobilität. In M. Groß (Ed.), *Handbuch Umweltsoziologie.* VS Verlag für Sozialwissenschaften.

Hadler, M., Schwarzinger, S., & Schweighart, M. (2019). *Forschungsbericht. Umweltspezifische Lebensrealitäten von MaturantInnen in der Steiermark.* Graz.

Hadler, M., Schweighart, M., & Wardana, R. (2021). *OeNB CO2-relevant environmental behavior.* Data will be available for free at the Austrian Social Science Data Archive. www.aussda.at; https://doi.org/10.11587/WQGMKY

Kollmuss, A., & Agyeman, J. (2002). Mind the gap: Why do people act environmentally and what are the barriers to pro-environmental behavior? *Environmental Education Research, 8*(3), 239–260.

Mairesse, O., Macharis, C., Lebeau, K., & Turcksin, L. (2012). Understanding the attitude-action gap: Functional integration of environmental aspects in car purchase intentions. *Psicológica, 33*, 547–574.

Moser, S. C., & Berzonsky, C. L. (2015). There must be more: Communication to close the cultural divide. In K. O'Brien & E. Selboe (Eds.), *The adaptive challenge of climate change.* Cambridge University Press.

Neugebauer, B. (2004). *Die Erfassung von Umweltbewusstsein und Umweltverhalten.* (ZUMA-Methodenbericht, 2004/07). Mannheim: Zentrum für Umfragen, Methoden und Analysen.

Preisendörfer, P. (1999). *Umwelteinstellungen und Umweltverhalten in Deutschland. Empirische Befunde und Analysen auf der Grundlage der Bevölkerungsumfragen „Umweltbewußtsein in Deutschland 1991–1998".* Springer Fachmedien.

Spezial Eurobarometer (2008). *Einstellungen der europäischen Bürger zur Umwelt. Bericht.* Spezial Eurobarometer 295/Wave 68.2—European Opinion Research Group EEIG.

Tanner, C., & Foppa, K. (1996). Umweltwahrnehmung, Umweltbewußtsein und Umwelt-verhalten. In A. Diekmann & C. C. Jaeger (Eds.), *Umweltsoziologie. Kölner Zeitschrift für Soziologie und Sozialpsychologie, Sonderheft 36/1996* (pp. 245–271). Westdeutscher Verlag.

CHAPTER 8

International Outlook and Conclusions

After developing an instrument for measuring climate-relevant behaviors, considering the factors that shape these behaviors as well as their determinants and obstacles in the Austrian context, this final chapter[1] discusses the same issues at an international level. It starts with an overview of the 2020 ISSP (International Social Survey Programme) questionnaire, which includes new questions related to housing, mobility, and diet. The ISSP fieldwork, however, is delayed due to the COVID-19 crisis, and we thus use data from the ECHOES project (Reichl et al., 2019) to show that our models also apply to other European countries. The section on the questionnaire development is followed by an analysis showing the different lifestyles from Chap. 6 at an international level and a brief consideration of the effects of the COVID-19 crisis on environmental attitudes and concerns. The analysis of the lifestyles shows that they can be found in other European countries as well. As for the COVID-19 crisis, we find a positive association between worries about COVID-19 and environmental attitudes. Yet, it is too early to assess if the pandemic has altered environmental concerns and behaviors permanently. The chapter concludes with a few final remarks.

---

[1] Lead author: Markus Hadler.

© The Author(s) 2022
M. Hadler et al., *Surveying Climate-Relevant Behavior*,
https://doi.org/10.1007/978-3-030-85796-7_8

8.1   INTERNATIONAL SURVEYS AND EXPLAINED VARIANCE

The previous chapters discussed the development of valid and reliable measures for climate-relevant behaviors in the Austrian context. The work associated with the development of these questions also informed the development of the International Social Survey Programme's (www.issp.org) 2020 Environmental Attitudes and Behavior questionnaire. The development of all ISSP questionnaires is led by a drafting group, which reviews current literature, proposes different questions to the general assembly, and also conducts pre-tests with the proposed items. The 2020 drafting group was convened by the Austrian representatives Markus Hadler and Markus Schweighart. The other members were the ISSP representatives from Chile, Spain, Sweden, and Taiwan, as well as the two external experts Malcolm Fairbrother and Axel Franzen. The final adoption of all questions took place in a democratic process during the 2019 ISSP meeting, in which all member countries have a single vote.

The final questionnaire includes 60 questions on environmental attitudes and behaviors and an additional comprehensive set of variables on socio-demographics. As for climate-relevant behaviors, items on transportation, housing, and food were included. The exact items are:

(a) "In the last twelve months, how many trips did you make by plane? Count outward and return journeys, including transfers, as one trip."

(b) "In a typical week, about how many hours do you spend in a car or another motor vehicle, including motorcycles, trucks, and vans, but not counting public transport? Do not include shared rides in buses, minibuses, and collective taxis."

(c) "In a typical week, on how many days do you eat beef, lamb, or products that contain them?"

(d) "How many rooms are there in your home (apartment or house)? Do not count any separate kitchens, bathrooms, garages, balconies, hallways or cupboards."

(e) Several background questions on the household structure.

Unfortunately, the fieldwork of ISSP was delayed due to the COVID-19 crisis and the data collection is still taking place in some countries. The international dataset is now supposed to be available in 2023 and our comparison in this concluding chapter, thus, uses data from the ECHOES project, which also included detailed questions on climate-relevant behaviors of Europeans. The questions are slightly different than what we had developed, but are close enough to get an idea of the overall picture. ECHOES, on the other hand, doesn't include many items on environmental attitudes and concerns. Furthermore, data was collected using existing online panels, whereas ISSP allows only random samples with postal or face-to-face recruiting. Yet, the ECHOES data allow us to test our assumptions at a basic level.

In Chap. 5 we concluded that the items on car usage, beef and lamb consumption, number of flights per year, the living space in square meters, as well as the number of household members are able to account for more than three quarters of the greenhouse gas (GHG) emissions. We used the most similar items from the ECHOES project and ran a regression for each country and noted the explained variance. Fig. 8.1 shows which percentage of the variance can be explained using the items "Q8_number_residents," "Q75_km_driver," "Q92_flights_private," "Q95_dwelling_size," and "Q106_diet" plus the question "Q104_green_provider," which asks whether or not the energy provider is a green provider. This question increased the explained variance by around ten percentage points for the majority of countries. Hence, this question could be a valuable addition to an international comparative survey in the European context.

We also tried to find out, which country-level characteristics are associated with a high level of explained variance. A correlation analysis at the aggregate level suggests the explanation is higher in countries which are more affluent ($r = 0.29$), have a larger share of green parties ($r = 0.36$), and where the public is more willing to make sacrifices for the environment ($r = 0.42$). Our instrument, hence, seems to be better applicable in more affluent societies and societies in which environmentalism is more widespread. Our instrument, thus, seems to be well suited to account for the GHG emissions in those societies and in those social groups that produce the most emissions. The UNEP emission report (2020) highlights that the top 1% of income earners produce around 15% of the emissions, and the top 10% of income earners (including the 1%) produce 48% of the emissions. These groups are well captured in the ECHOES survey. The ISSP data then will show if and how strongly the emissions-relevant behaviors also differ within developing countries.

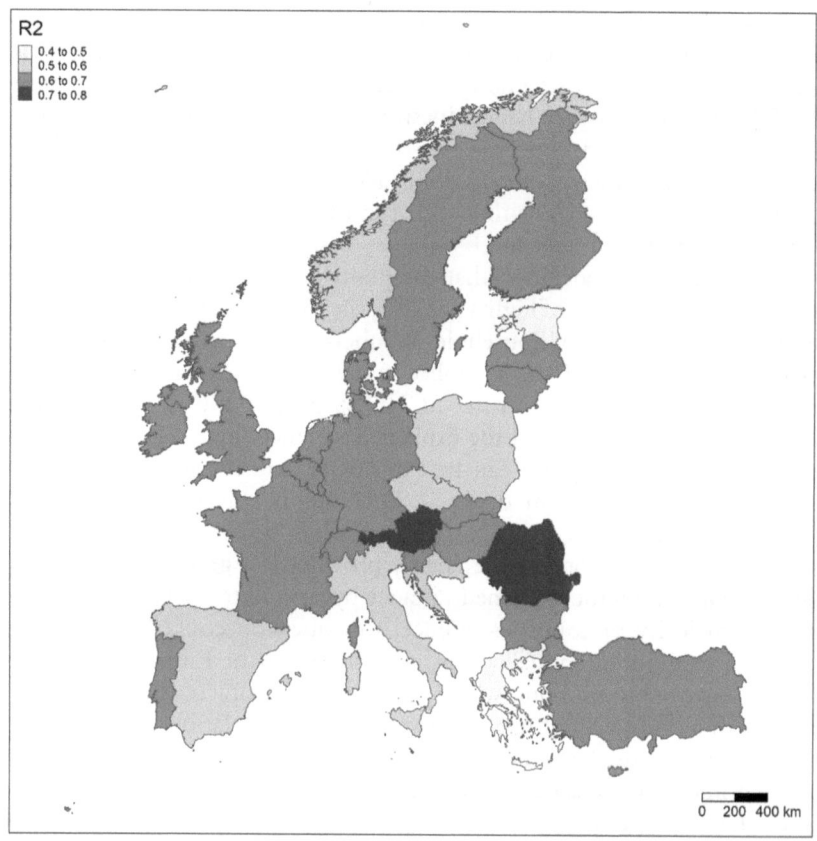

Fig. 8.1 Explained variance of total GHG emissions. Explained variance (r^2) in regression with "Q8_number_residents," "Q75_km_driver," "Q92_flights_private," "Q95_dwelling_size," "Q106_diet," and "Q104_green_provider" as independent variables. (Source: Calculation by DN Bird based on ECHOES data [Reichl et al., 2019])

8.2   Lifestyles in an International View

Chapter 6 showed that the Austrian population can be clustered into five distinctive groups based on their energy demand. Using data from the ECHOES project, we now test whether these groups can also be found when using GHG emissions instead of the energy demand and, if so, how large they are in other European countries. Subsequently, we look into the factors that shape their occurrence. These analyses, however, are limited to the country level and thus rest on a maximum of 31 observations.

The analyses based on GHG emissions result in one to two additional lifestyle groups depending on the clustering criteria used. These additional groups are a cluster of "averages" and a cluster with "low consumption in food." For the sake of comparability to Chap. 6, we use only the four groups in this international outlook, which occur in both analyses, that is, in the GHG emission and the energy demand approach. Furthermore, we omit the group with a high demand in leisure and information from Chap. 6, as this group splits into new groups when changing to a GHG approach.

The four lifestyles we consider in detail are Homers, Travelers, Savers, and Consumers. Homers are characterized by high consumption in the area of housing, Travelers by high mobility in terms of car usage and air travel, and Consumers by high goods usage. Savers is the group that is below average in all categories. Figure 8.2 provides an overview of their prevalence in different European countries. It shows that Savers and Travelers are more common than Homers and Consumers in all countries. The former groups reach a prevalence of more than 30% of the population in some countries, whereas Consumers are commonly around 10% and Homers often below this figure. Furthermore, the occurrences of Consumers and Savers are negatively correlated, indicating that they represent opposed patterns.

Alongside the prevalence of these lifestyles, we also consider the factors that shape their extend and occurrence. The literature points to an association between environmental impact and affluence, population, attitudes, environmental state, and political factors as pointed out in Chap. 2. We hence correlated the size of the lifestyle clusters in the European countries with related country characteristics. In particular, we included the level of affluence (GDP per capita) and human development (HDI), population density (people per square km), age structure (% of population above 65 years) and urbanization (% of the population living in cities), the state of the environment in terms of water quality, GHG emission per capita, biomass available per capita; environmental attitudes and concerns in terms of percentage of the population that worries about the climate and their willingness to do something for the environment, as well as political factors such as the comparative magnitude of the green party in the last national election, expenditures on environmental protection (% of GDP), and the protected land areas (% of total area). Finally, we also considered

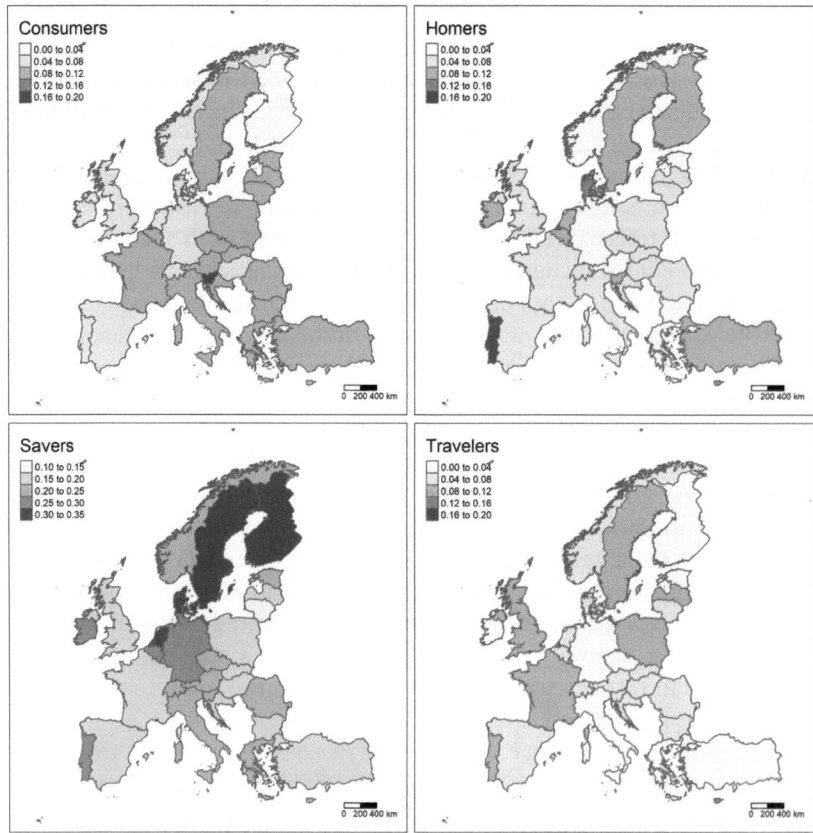

Fig. 8.2 Lifestyles across Europe. (Source: Calculation by DN Bird based on ECHOES data [Reichl et al., 2019])

the mobility indicators "cars per 1000 people" and "railroad km per head." Data were derived from Eurostat, the European Social Survey, the Manifesto Project, OECD, UNDP, the World Bank, and the Yale Environmental Performance Index.

Considering the correlation between these indicators and the frequency of the lifestyles shows that indicators of development are positively correlated with the frequency of Savers and negatively with Consumers, whereas there are only very weak and inconsistent associations with the percent of Homers and Travelers. Demographic factors have the strongest effects on

Homers. Homers are more common in densely populated and more urban countries. The state of the environment in terms of GHG per capita, the available biomass per capita, water quality, and so on do not show any strong correlation with the exception of available biomass, which is correlated with Consumers and somewhat with Travelers. There is, however, no obvious substantive interpretation for this correlation.[2] Worries about the climate are associated with a large number of Homers and the readiness to make sacrifices for the environment is correlated with the frequency of Savers. As for the political institutionalization of environmental topics, larger shares of green parties are associated with fewer Consumers, higher expenditures on environmental measures with larger shares of Savers, and the percentage of protected areas of land with a larger number of Consumers.

Interpreting these associations from a lifestyle view shows that Consumers are more common in countries that are less affluent, and have a lower environmental performance and a less influential green party. In the Austrian context of Chap. 6, this group was characterized by buying lots of fashion and electronics—in other words expressing a more materialistic lifestyle. Using Inglehart's view on the development of post-materialism, we would have indeed expected to find them more often in less affluent countries. The findings at the micro-level and the international outlook are thus in line. The Homers can be found in dense and urban countries in our international outlook. In the Austrian context, the analysis shows a trend toward single-person and smaller households and older respondents. There is however no correlation between age structure and this lifestyle in the European context. This lifestyle, hence, seems to be more common with ongoing aging and urbanization within and between countries. Again, the micro-level interpretation and the international outlook are mostly aligned. The picture is more complex as far as Savers are concerned. This lifestyle is more common in affluent countries and societies where the willingness to make sacrifices is high and environmental protection is well established. Interestingly, this group was characterized by lower status and limited labor market integration within the Austrian context. The national finding and the international outlook lead to opposite interpretations. Travelers, finally, are not linked to any specific societal characteristics considered so far. The national finding was that

---

[2] Land use and biomass are also associated with high unemployment and lower affluence. Hence, the associations between land use/biomass and lifestyles might be spurious.

men and better educated individuals are typical for this group. There is a positive correlation between rail passengers (per head) and Travelers, but not with cars per 1000 inhabitants. The latter, however, is correlated with the frequency of Savers (probably due to development correlation).

8.3   COVID-19 CRISIS AND IMPACT

As pointed out in the introduction, the advent of the COVID-19 crisis brought a sudden end to the attention given to the climate crisis. Yet, the restrictions imposed by the different governments to address the COVID-19 crisis led to a reduction in GHG emissions. The UNEP (2020) emissions report shows in this regard that the emissions dropped significantly in the areas of ground transportation, power, industry, and aviation. These reductions, however, occurred due to a mandated change and were not based on the voluntary behavioral changes of individuals. It is thus unclear how the COVID-19 crisis will affect the underlying environmental attitudes and behaviors in the long run.

During the first wave of the COVID-19 crisis in 2020, Austria took part in the Values in Crisis Study (VIC, 2021), which was initiated by researchers from the World Values Survey (WVS, www.worldvaluessurvey. org). This survey includes questions from the WVS and COVID-19 related items. At the time of this publication, merged data was available for Austria, Brazil, China, Colombia, Germany, Georgia, Greece, Italy, Japan, Kazakhstan, Maldives, Poland, South Korea, Sweden, and the United Kingdom. The Austrian data (Aschauer et al., 2020) also included a question on the "willingness to make sacrifices for the environment," which according to Mayerl and Best (2019) reflects the behavioral dimension of the tripartite classification of Maloney and Ward's (Maloney & Ward, 1973) ecology scale, is part of Dunlap and Jones' (2002) environmental concern, and can be seen as a behavioral intention variable that fits Ajzen and Fishbein's (1980) attitude-behavior model.

In the Austrian context, we were able to show that the willingness to make sacrifices for the environment in terms of paying higher prices and taxes as well as accepting reductions in the standard of living dropped during the COVID-19 crisis (Klösch et al., 2021). The international VIC dataset includes unfortunately only a single item on environmental attitudes. Respondents are asked on a 6-point scale whether the description "She/he strongly believes that people should care for nature. Looking after the environment is important to her/him" is "very much like me" to

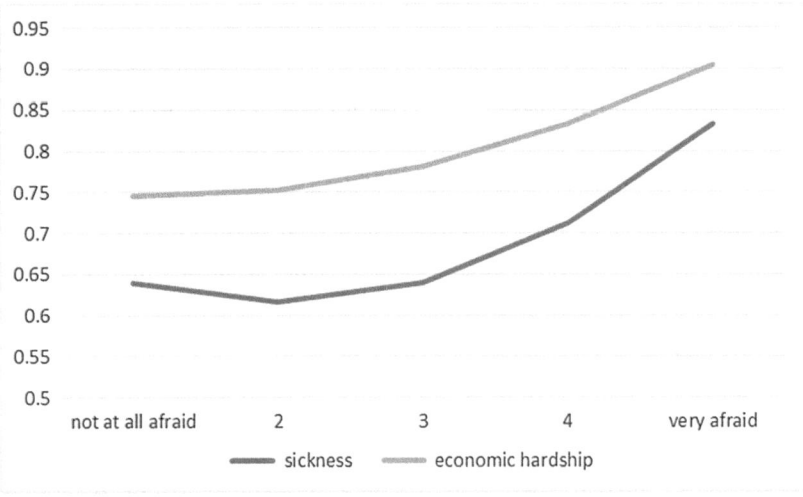

Fig. 8.3 Association between worries about economic and health impact of COVID-19 and the self-perception as person who acts for the environment. (Source: Regression using VIC [2021] data)

"not at all like me." We used this item in a regression considering education, age, income, gender, left-right scale, and country of residence as controls and worries about economic hardship and health issues due to COVID-19 as independent variables.

Figure 8.3 shows the average effects across all countries, based on the regression mentioned before. The vertical axis displays the predicted self-perception with a higher value indicating a more environmentally friendly self-perception. The horizontal axis displays the strength of the COVID-19 fears with an increasing worry from the left to the right. The relationship between worries about COVID-19 and the self-perception as an environmentalist is slightly curvilinear. Overall, respondents who are worried about COVID-19 also care about the environment. A possible interpretation is that we observe a general underlying notion of concern, which is in line with the benevolence and universalism dimension of Schwartz's value theory (2012).

These findings indicate an increasing "environmentalism" due to COVID-19 fears. Yet, the VIC question on the self-perception must not be considered equal to a measure of environmental behaviors. As pointed out before, we noticed a declining willingness to make sacrifices for the environment during the COVID-19 crisis in the extended Austrian data-set (Klösch et al., 2021). Furthermore, we also found that in particular respondents who are worried about the economic impact of COVID-19 are less willing to make sacrifices for the environment.[3] This is not surprising, assuming that individuals who struggle economically most likely do not have the means to spend more money on the environment. The willingness item, however, is not included in the international dataset and we thus cannot test it empirically at an international comparative level. Future research will have to look into the long-term effects of the COVID-19 crisis on climate-relevant behaviors and related concerns.

8.4   Overall Impact and Individual Actions

So far, we have considered the measurement of GHG emissions, its international applicability, and also the effects of the COVID-19 crisis on environmental attitudes and behaviors. The most pressing questions concerning our book in terms of societal resilience is how the climate crisis can be tackled and how to reduce GHG emissions to a level that is in line with the Paris goals. We can recur to the IPAT equation (Rosa et al., 2015; Harper, 2015) for this purpose and consider the factors population "P," affluence "A," and technology "T" when assessing the factors that shape the impact "I," that is, GHG emissions in our case.

The first main factor is population size. First, the total emissions of a society depend strongly on the size of population, even when controlled for technology and affluence. In this regard, larger societies such as China and the United States emit a large amount of GHG. The 2019 figures are 14 $GtCO_2e$ for China and 6.6 $GtCO_2e$ for the United States (UNEP, 2020, p. 5). The size of the impact of single individuals, on the other hand, is estimated more accurately when considering the per-capita output. Here, the figures are 9.7tCO_2e for China and 20.0tCO_2e for the United States (ibid.). Given that the total number of emissions counts, reductions have to be particularly large in growing societies.

The second factor, affluence, points to the consumption patterns. The comparison of the per-capita GHG output in the United States and China showed that the former consumes much more per head, which points to the unequal distribution of consumption and GHG emissions. According

---

[3] Interestingly, also respondents, who did not worry at all about COVID-19, were less willing to make a sacrifice for the environment.

to the UNEP (2020), the richest 10% of the global income earners emit around 48% of the GHG and the poorest 50% only 7% of all GHG. To balance increasing development, population growth, affluence, associated increasing emissions, and the overall need to reduce the total emissions of GHG evokes questions of social justice and fairness. Is it, for example, right to demand that developing societies not become more affluent, so that the affluent societies have to change less? We doubt it.

Technology, the third factor, refers to production processes and also to the combustion material used. Theories on ecological modernization proposed that the impact would be smaller once technology advances. Results, however, are mixed (Harper, 2015). As for climate change, ideas such as GHG storage, shielding the planet from sun rays are the most extreme ideas. More realistically, the replacement of fossil fuels by renewable energy sources, such as oil-based heating systems with heat pumps or hydrogen instead of petrol, is plausible. Yet, we can run into a Jevon's problem in the sense that an increasing use of materials offsets the increase in efficiency, which is also related to the previous paragraphs on population growth and increasing affluence for large parts of the world's population.

The IPAT formula offers a broad overview of the different dimensions, but does not aim to provide any specific guidelines and ideas for individuals. The current UNEP emission report includes an entire chapter that focuses on how to shape individual action (UNEO, 2020, pp. 70–73). It distinguishes between (a) the social and contextual circumstances such as media, social norms, social movements, and so on; (b) structural circumstances such as policies, infrastructures, supply chains, and others; and (c) personal and immediate circumstances such as knowledge, attitudes, habits, and so on. All of these three circumstances shape lifestyles and, in turn, lifestyles and actions of individuals also shape these circumstances.

Using these divisions as a backdrop, the UNEP report highlights a few mechanisms that may change behaviors. It points out that, first, incentives, information, and choice provision do work to a certain extent, but larger sustainable change can occur only by changing social norms and the options available (UNEP, 2020, p. 71). Restrictions and laws also work well, but depend on the acceptance by the population. Second, the available infrastructure can be addressed. Here, the availability of public transport and suburbanization, and so on play a role. Third, social influence, in the sense that a single person can influence their social environment, is mentioned. Installing a photovoltaic system on a house, for example, might entice other neighbors to do the same, initiating social change.

Fourth, citizen participation in social movements and inclusion in politics also empowers individuals and might increase the likelihood of changing to more climate-friendly behaviors. Finally, the report also points to disrupting habits—to challenge routine behaviors. The latter occurred prominently during the COVID-19 crisis.

The findings presented in this book are related to these points. Chapter 5 showed that the total GHG emissions of our respondents are related to their place of residency, that is, urban, suburban or rural, income, and age. These findings are similar to previous studies which found a strong influence of these factors on consumption (Poortinga et al., 2004). As for attitudes and concerns, the willingness to make sacrifices for the environment had significant effects, whereas attitudes such as fatalism were not significant. Of course, we need to bear in mind that our analysis remained at an overall level and that attitudes might become significant, when comparing individuals in similar life circumstances. Furthermore, previous studies already pointed out that the effects of attitudes are stronger on the intention to change behaviors (Abrahamse & Steg, 2009). Overall, this first analysis underscored the influence of the context and hence points to infrastructural measures as a lever for change.

The subsequent chapter considered the different lifestyles of individuals and their energy consumption in six different areas of social life (see also Schwarzinger, Bird, & Hadler, 2019a, Schwarzinger, Bird, & Skjølsvold, 2019b). The lifestyles of Travelers and Homers turned out to use the most energy of all groups. Considering these two areas, policy measures aiming at an effective reduction of energy demand will have to look into replacing fossil fuels, increasing the efficiency of heating systems and means of transportation, or into altering behaviors. The climate crisis, however, cannot be solved only by considering these groups. Firstly, they are rather small in size. Secondly, even the group with the lowest consumption—the Savers—is exceeding the allotted GHG emissions by far. Policy measures, thus, need to target even very frugal Consumers in the Western societies.

Chapter 7 considered the perceived obstacles to act in an environmentally sound way. The interviewees frequently mentioned a lack of information and pointed to the limited choices in goods and public transportation, especially in the rural areas. The presence of a willingness to do something is bound by the options. Interviewees suggested changes at the structural and political level, that is, that a change in market regulation is desired, especially with regard to offering and promotion of eco-friendly alternatives, but also pointed to increasing education, information dissemination, and awareness raising among the population in order to counteract intra-personal barriers. These suggestions are also in line with findings from the "Model of Pro-Environmental Behavior" by Kollmuss and Agyeman (2002), that is, the importance of structural conditions as well as routines and habits for pro-environmental behavior, and with the importance of situational factors in Barr's "Framework of Environmental Behavior" (Barr, 2004, 2006). However, we also need to consider that certain decisions lie outside of routine actions, such as buying a car or moving to a larger place. Here, zoning and policies that limit choices might be advantageous as well.

8.5   Concluding Remarks

Our book started with the idea of developing a survey that allows researchers to measure a respondent's greenhouse gas emissions in a concise manner. We were successful in the sense that we were able to identify a few questions on mobility, housing, and diet that are able to account for more than three quarters of a respondent's emissions. While these three areas were known beforehand as GHG-intensive areas, the specific contribution is that we offer survey researchers specific guidance on this task. Furthermore, while focusing initially on Austria, we were able to show that these items also work in other European countries.

In terms of explaining the GHG of respondents, we confirmed, on the one hand, that the standard models of environmental behavior apply and that the overall GHG output is heavily influenced by context, situational variables, and socio-demographics. Our analyses, however, pointed to several new directions that need to be considered. First, we identified specific lifestyles and patterns of consumption that showed that some social groups have a strong impact in only one or two areas and that some respondents are below average in all areas. Climate policies need to consider these specific patterns. Second, we also showed that the use of multiple methods and data sources allow for a more holistic picture. The initial analysis was based on a survey. Based on the information derived from this survey, qualitative interviews were conducted with individuals who show a gap between their attitudes and behaviors. Our next step, then, will be to extend our analyses beyond the European context. Finally, our different analyses and findings showed that there is need for a variety of different support measures and options in order to facilitate climate-friendly behaviors. Yet, we must not forget that there are also other social and environmental problems that need to be addressed at the same time.

REFERENCES

Abrahamse, W., & Steg, L. (2009). How do socio-demographic and psychological factors relate to households' direct and indirect energy use and savings? *Journal of Economic Psychology, 30*(5), 711–720.

Ajzen, I., & Fishbein, M. (1980). *Understanding attitudes and predicting social behavior.* Prentice-Hall.

Aschauer, W., Seymer, A., Prandner, D., Baisch, B., Hadler, M., Höllinger, F., & Bacher, J. (2020). *Values in crisis Austria (SUF edition).* Retrieved January 11, 2021, from https://doi.org/10.11587/H0UJNT, AUSSDA, V1, UNF:6:FkbX1WYV0re1Mq1Ae1ICTA== [fileUNF].

Barr, S. (2004). Are we all environmentalists now? Rhetoric and reality in environmental action. *Geoforum, 35,* 231–249.

Barr, S. (2006). Environmental action in the home: Investigating the 'value-action' gap. *Geography, 91*(1), 43–54.

Dunlap, R. E., & Jones, R. E. (2002). Environmental concern: Conceptual and measurement issues. *Handbook of Environmental Sociology, 3*(6), 482–524.

Harper, C. (2015). *Environment and society: Human perspectives on environmental issues (2-downloads).* Routledge.

Klösch, B., Wardana, R., & Hadler M. (2021). *The negative impact of the COVID-19 crisis on the willingness to act for the environment.* Results from a survey of Austrians during the COVID-19 pandemic. Unpublished manuscript.

Kollmuss, A., & Agyeman, J. (2002). Mind the gap: Why do people act environmentally and what are the barriers to pro-environmental behavior? *Environmental Education Research, 8*(3), 239–260.

Maloney, M. P., & Ward, M. P. (1973). Ecology: Let's hear from the people: An objective scale for the measurement of ecological attitudes and knowledge. *American Psychologist, 28*(7), 583–586. https://doi.org/10.1037/h0034936

Mayerl, J., & Best, H. (2019). Attitudes and behavioral intentions to protect the environment: How consistent is the structure of environmental concern in cross-national comparison? *International Journal of Sociology, 49*(1), 27–52.

Poortinga, W., Steg, L., & Vlek, C. (2004). Values, environmental concern, and environmental behavior: A study into household energy use. *Environment and Behavior, 36*(1), 70–93. https://doi.org/10.1177/0013916503251466

Reichl, J., Cohen, J., Kollmann, A., Azarova, V., Klöckner, C., Royrvik, J., Vesely, S., Carrus, G., Panno, A., Tiberio, L., Fritsche, I., Masson, T., Chokrai, P., Lettmayer, G., Schwarzinger, S., & Bird, N. (2019). *International survey of the ECHOES project.* Dataset. Zenodo. https://doi.org/10.5281/zenodo.3524917

Rosa, E. A., Rudel, T. K., York, R., Jorgenson, A. K., & Dietz, T. (2015). The human (anthropogenic) driving forces of global climate change. *Climate Change and Society: Sociological Perspectives, 2,* 32–60.

Schwartz, S. H. (2012). An overview of the Schwartz theory of basic values. *Online Readings in Psychology and Culture, 2*(1), 2307–0919. https://doi.org/10.9707/2307-0919.1116

Schwarzinger, S., Bird, D. N., & Hadler, M. (2019a). The 'Paris lifestyle'—Bridging the gap between science and communication by analysing and quantifying the role of target groups for climate change mitigation and adaptation: An interdisciplinary approach. In *Addressing the challenges in communicating climate change across various audiences* (pp. 375–397). Springer.

Schwarzinger, S., Bird, D. N., & Skjølsvold, T. M. (2019b). Identifying consumer lifestyles through their energy impacts: Transforming social science data into policy-relevant group-level knowledge. *Sustainability, 11*(21), 6162.

United Nations Environment Programme. (2020). *Emissions gap report 2020.* Nairobi. https://www.unep.org/emissions-gap-report-2020

VIC. (2021). *Values in crisis survey.* Initiated by C. Welzel, K. Boehnke, J. Delhey, F. Deutsch, J. Eichhorn, & U. Kühnen. https://www.worldvaluessurvey.org/WVSNewsShow.jsp?ID=416&ID=416

Questions Included in First Wave of the OeNB Study (Hadler et al. 2021)

Socio-demographic Variables

These include year of birth, highest completed school education, number of people living permanently in household (number of children, number of adults), monthly net income, and monthly household net income. Since the survey data were collected face-to-face, the questionnaire also included a number of questions that had to be answered by the interviewees when the interview was held in the respondent's home, including gender, name of district/municipality, amount of electronics within the home (a lot of new electronics, average equipment, economical equipment, no electronics at all except cell phone), window quality assessment (good quality, standard, standard to bad, bad quality), and a Proband code.

Environmental Attitudes and Personal Environmentally Relevant Behavior (PEBs)

- In general, how concerned are you about the environment? (Not worried at all–Very worried; 5 steps)
- In the next few questions, we are interested in your attitudes toward environmental issues. For each of the following statements, please indicate to what extent you agree or disagree. (Strongly agree–Strongly disagree at all; 5 steps)

© The Author(s) 2022
M. Hadler et al., *Surveying Climate-Relevant Behavior*,
https://doi.org/10.1007/978-3-030-85796-7

- For someone like me, it is simply difficult to do much for the environment.
- I do what is right for the environment even if it costs me more time and money.
- There are more important things to do in life than protecting the environment.
- It is useless to do my part for the environment as long as others do not behave in the same way.
- Many assertions about the danger to the environment are exaggerated.
- It is difficult for me to judge whether my lifestyle benefits or harms the environment.
- Modern science will solve our environmental problems with little change in our way of life.
- Today we worry too much about the future and too little about prices and jobs.
- Almost everything we do in our modern world harms the environment.
- People worry too much that human progress is damaging the environment.
- To protect the environment, Austria needs economic growth.
- Economic growth always harms the environment.

- To what extent would you personally find it acceptable for you to ... (Very acceptable–Very unacceptable; 5 steps)
  - ... pay much higher prices to protect the environment.
  - ... pay much higher taxes to protect the environment.
  - ... sacrifice your standard of living to protect the environment.
- How often do you do the following things? [Always–Never; 4 steps]
  - Separate valuable materials from your waste, such as glass, metal, plastic, paper, and so on, for reuse (recycling).
  - Buy fruit and vegetables that have not been treated with pesticides or chemicals.
  - Limit driving for the sake of the environment.
  - Reduce energy and fuel consumption at home for the sake of the environment.
  - Save or reuse water for the sake of the environment.
  - For the sake of the environment avoid buying certain products.

HOUSING—BUILDING INFORMATION

- How many apartments are there in the building you live in?
- How many exterior walls does your apartment have?
- When was the building you live in built?
- How many square meters of (interior) living space (without basement) does your apartment/house have?
- How is your apartment/house thermally insulated?
  - External wall insulation, roof insulation, basement ceiling insulation, windows well insulated
- How would you yourself estimate the average quality of your windows?
- Type of building
  - Single-family house, semi-detached house/terraced house, block of flats/high rise

HOUSING—HEATING AND HEATING BEHAVIOR

- What proportion of your living space is heated during the heating season?
- What is the main heating system used in your apartment?
- Is a supplementary heating system used in your apartment?
- Which main energy source is used to heat your apartment?
- If a heating bill is available:
  - Building/property: total kWh, total (consumption) units
  - Single apartment: (consumption) units or kWh
- If no heating bill is available or information is not given:
  - What is the amount of the energy source you consume per year? (Indication in liters, m³, kg, solid cubic meters, room meters)
  - Proof or estimate?
- What are the monthly costs for heating in your apartment/house?
- What is the temperature of the room where you spend most of the day during the heating season?
- Compared to other apartments/houses in Austria, how would you rate the room temperature in your apartment during the heating season?
- How often do you reduce the room temperature during the heating season … (… at night? … if you leave the apartment for more than four hours? … if you leave the apartment for a day? … if you leave the apartment for more than a day?)

- If you have an energy certificate, please indicate the heating demand in kWh/m².

HOUSING—POWER CONSUMPTION

- How often are the following devices used in your household? (Cast iron stove, ceramic stove top, induction stove, gas stove, oven, microwave, dishwasher, washing machine, tumble dryer)
- How many hot meals do you personally eat on average per week? (also away from home)
- On average, how many people in your household are being cooked for?
- How many of the following appliances are there in your household in total and how many hours a day do you personally use these appliances on average? (Television [≤40 inch], television [>40 inch] or beamer, stereo/home cinema system, stand-up PC, laptop, air conditioning [use related to summer])
- Choose the answer that best suits your personal use of consumer electronics (PC, notebook, TV, Hi-Fi equipment) excluding smartphones. (I use little, I use less than most, I use average, I use more than most, I use very intensively)
- How high was your electricity consumption last year?
  - Proof or estimate?
  - What are your monthly electricity costs?
  - Alternative: Electricity meter reading

HOUSING—WATER TREATMENT AND WATER CONSUMPTION

- Which of the following water heating techniques do you use? (Instantaneous water heater/hot water boiler, storage tank [boiler], district heating, hot water heat pump, solar heating system)
- How often do you shower on average per week?
- How long do you shower on average?
- How many baths do you take per month on average? (Number of baths per month? How many of these baths are full baths?)

MOBILITY—INDIVIDUAL MOTORIZED MEANS OF TRANSPORT

- How many of the following motorized vehicles does your household own? (Car, motorcycle, moped/scooter, bus/camper van/tractor, or similar)
- How many kilometers have you covered in the last 12 months with a car? (as driver * in and/or passenger * in)
- What is the mileage of the car?
- On average, how many hours per week do you spend on the road in a car? (as driver * in and/or passenger * in)
- How often are you alone in the vehicle?
- How many people are usually in the car when you are not alone? (as driver * in and/or passenger * in)
- What is the type of fuel used by your most common car?
- How many liters does your most-used car consume on average per 100 kilometers?
- How many hours per week on average do you travel by public transport (train, bus, streetcar, metro)?

MOBILITY—FLIGHT BEHAVIOR

- How many flights have you taken in the last 12 months privately or professionally in a passenger aircraft (count each single flight separately)? (Number of short and medium distance [up to 3000 km or 3.5 h flight time], number of long distance [more than 3000 km or 3.5 h flight time])
- In the last 12 months, how many hours have you spent on private or business flights (times between departure and landing)? (Flying hours private, flying hours professional)
- What flights have you taken in the last 12 months? (Starting point, destination)
- Which of the following statements is most likely to apply to your flight behavior? (I fly abroad several times a year [also long-distance flights], I fly abroad several times a year [mainly short distance flights], I fly abroad about once a year, I fly abroad once every few years, I almost never fly abroad, I never fly)

DIET—CONSUMPTION OF ANIMAL PRODUCTS AND WASTE

- How often do you eat the following foods? (Sausage products, beef and veal, pork, poultry, lamb, fish and seafood, cheese/eggs, ready meals, frozen products)
- Choose the answer that best describes your eating habits? (Meat in most meals, meat in some meals, meat very rarely, no meat but fish, vegetarian, vegan)
- What is the approximate amount of food that is thrown away in your household (in percentage)?
- On average, how often do you eat meals in restaurants and the like (including fast food, delivered meals, etc.)?

CONSUMPTION—PURCHASING BEHAVIOR CLOTHING, ELECTRONIC DEVICES, LEISURE TIME BEHAVIOR

- Choose the answer that most closely corresponds to your purchasing behavior of electronic items (PC, notebook, tablet, smartphone, TV, game console, Hi-Fi). (I don't need most of it, I take care to use it for a long time and replace electrical items only when they break, I buy new equipment from time to time even if the old one is not broken, I buy new equipment regularly, I make sure I always have the latest technology)
- Describe how you deal with major household investments (e.g., refrigerator, washing machine, kitchen, TV). (Few and modest new acquisitions, thoughtful purchase of durable products, average, much but rather cheap, gladly the newest, full luxury equipment)
- Choose the answer that best suits your approach to clothing. (Very modest, long use, average, often new clothes, always in the latest style)
- How many of the following things have you bought or received as gifts in the last 12 months (with the exception of second-hand clothing)? (Shoes, shirts/tops/blouses, trousers/skirts, pullovers/dresses/blazers, jackets/coats, CDs/DVDs/vinyl/Blu-Rays, books)
- On average, how often do you buy the following items new? (Smartphone/mobile phone, PC/laptop, television, car, bicycle, ski/snowboard)

- On how many days in the last 12 months have you visited the following facilities? (Cinema/theater/opera/soccer stadium, ski resort, other amusement parks, hotel, apartment/bed and breakfast/ youth hostel)
- Choose the answer that best describes your hobbies and leisure activities. (Very little equipment and infrastructure needed [e.g., board games, reading], little equipment and infrastructure needed [e.g., music, hiking, cycling], moderate amount of equipment and infrastructure needed [e.g., video games, photography], some equipment and infrastructure needed [e.g., skiing, team sports], much equipment and infrastructure needed [e.g., motor sports])

Reference

Hadler, M., Schweighart, M., & Wardana, R. (2021). *OeNB CO2-relevant environmental behavior.* Data will be available for free at the Austrian Social Science Data Archive. www.aussda.at; https://doi.org/10.11587/WQGMKY

Index[1]

---

[1] Note: Page numbers followed by 'n' refer to notes.

© The Author(s) 2022
M. Hadler et al., *Surveying Climate-Relevant Behavior*,
https://doi.org/10.1007/978-3-030-85796-7